普通高等教育“十三五”规划教材

服务外包产教融合系列教材

主编 迟云平 副主编 宁佳英

Android
游戏开发

主 编 杜 剑

副主编 罗 林 李俊琴 刘志伟

华南理工大学出版社
SOUTH CHINA UNIVERSITY OF TECHNOLOGY PRESS
·广州·

图书在版编目(CIP)数据

Android 游戏开发/杜剑主编. —广州：华南理工大学出版社，2017.5(2018.7 重印)
(服务外包产教融合系列教材/迟云平主编)
ISBN 978-7-5623-5241-9

Ⅰ.①A… Ⅱ.①杜… Ⅲ.①移动终端-游戏程序-程序设计-教材
Ⅳ.①TN929.53 ②TP311.5

中国版本图书馆 CIP 数据核字(2017)第 069066 号

Android 游戏开发
杜 剑 主编

出 版 人：卢家明
出版发行：华南理工大学出版社
(广州五山华南理工大学 17 号楼，邮编 510640)
http://www.scutpress.com.cn E-mail:scutc13@scut.edu.cn
营销部电话：020-87113487 87111048（传真）
总 策 划：卢家明 潘宜玲
执行策划：詹志青
责任编辑：刘 锋 袁 泽
印 刷 者：佛山市浩文彩色印刷有限公司
开 本：787mm×1092mm 1/16 **印张**：12.75 **字数**：300 千
版 次：2017 年 5 月第 1 版 2018 年 7 月第 2 次印刷
印 数：1001～3000 册
定 价：28.50 元

“服务外包产教融合系列教材”
编审委员会

总　序

发展服务外包，有利于提升我国服务业的技术水平、服务水平，推动出口贸易和服务业的国际化，促进国内现代服务业的发展。在国家和各地方政府的大力支持下，我国服务外包产业经过10年快速发展，规模日益扩大，领域逐步拓宽，已经成为中国经济新增长的新引擎、开放型经济的新亮点、结构优化的新标志、绿色共享发展的新动能、信息技术与制造业深度整合的新平台、高学历人才集聚的新产业，基于互联网、物联网、云计算、大数据等一系列新技术的新型商业模式应运而生，服务外包企业的国际竞争力不断提升，逐步进入国际产业链和价值链的高端。服务外包产业以极高的孵化、融合功能，助力我国航天服务、轨道交通、航运、医药、医疗、金融、智慧健康、云生态、智能制造、电商等众多领域的不断创新，通过重组价值链、优化资源配置降低了成本并增强了企业核心竞争力，更好地满足了国家“保增长、扩内需、调结构、促就业”的战略需要。

创新是服务外包发展的核心动力。我国传统产业转型升级，一定要通过新技术、新商业模式和新组织架构来实现，这为服务外包产业释放出更为广阔的发展空间。目前，“众包”方式已被普遍运用，以重塑传统的发包/接包关系，战略合作与协作网络平台作用凸显，从而促使服务外包行业人员的从业方式发生了显著变化，特别是中高端人才和专业人士更需要在人才共享平台上根据项目进行有效整合。从发展趋势看，服务外包企业未来的竞争将是资源整合能力的竞争，谁能最大限度地整合各类资源，谁就能在未来的竞争中脱颖而出。

广州大学华软软件学院是我国华南地区最早介入服务外包人才培养的高等院校，也是广东省和广州市首批认证的服务外包人才培养基地，还是我国服务外包人才培养示范机构。该院历年毕业生进入服务外包企业从业平均比例高达66.3%以上，并且获得业界高度认同。常务副院长迟云平获评2015年度服务外包杰出贡献人物。该院组织了近百名具有丰富教学实践经验的一线教师，历时一年多，认真负责地编写了软件、网络、游戏、数码、管理、财务等专业的服务外包系列教材30余种，将对各行业发展具有引领作用的服务外包相关知识引入大学学历教育，着力培养学生对产业发展、技术创新、模式创新和产业融合发展的立体视角，同时具有一定的国际视野。

当前，我国正在大力推动"一带一路"建设和创新创业教育。广州大学华软软件学院抓住这一历史性机遇，与国家发展和改革委员会国际合作中心合作成立创新创业学院和服务外包研究院，共建国际合作示范院校。这充分反映了华软软件学院领导层对教育与产业结合的深刻把握，对人才培养与产业促进的高度理解，并愿意不遗余力地付出。我相信这样一套探讨服务外包产教融合的系列教材，一定会受到相关政策制定者和学术研究者的欢迎与重视。

借此，谨祝愿广州大学华软软件学院在国际化服务外包人才培养的路上越走越好！

国家发展和改革委员会国际合作中心主任

2017年1月25日于北京

前　言

如今市场上的移动操作系统众多，除了如日中天的iOS、Android，还有微软寄予厚望的Windows Phone、明日黄花的Symbian、新兴潜力无限的LiMo以及有固定拥趸的BlackBerry OS，等等。从美国市场占有份额看，2010年排第一位的Android已占了40%，并有连年上升的趋势。在中国使用Android手机的客户数也遥遥领先于其他平台。除了Google原生操作系统之外，还有众多公司基于Android进行二次开发的操作系统，如联想乐OS、阿里云OS、百度易等等。可以说，Android凭借开源免费的优势已经占据了领先的地位。随着众多手机厂商乃至运营商加入Android阵营，Android系统的应用将越发广泛并逐步走向成熟。

在手机众多应用当中，游戏所占比例一直排在首位，而随着游戏开发行业的发展与成熟，游戏项目外包已经成为行业中常见的开发方式。近几年，大量的游戏外包公司成立，外包的项目数量也在迅速增长。游戏服务外包的持续火热与网络游戏产业的迅猛发展有密切的关系。2015年中国网络游戏市场规模已达1300多亿元，如此巨大的市场吸引了众多企业参与其中。如何更快推出游戏应用软件以占领市场，是游戏开发企业必须考虑的问题，因此，将游戏开发中的非核心业务外包成为游戏开发企业的一个可能的选择。

目前我国出版的计算机图书中关于游戏开发的书籍并不多，关于Android游戏开发的更是寥寥无几，因此我们编写本书，希望能弥补这一不足。

本书阐述在Android平台上进行游戏开发所需技术，希望给渴望加入Android游戏开发行列的读者一些帮助。本书需要读者有一些Java语言开发的基础。

本书主要分为三篇：第一篇Android入门，带领大家了解Android平台的特点及今后发展的趋势，并帮助读者建立自己的第一个 Android 项目。第二篇 Android 游戏开发基础，涵盖 Android 游戏开发的基本技术点，如图形、声音、交互、网络、数据处理等。第三篇 Android 游戏开发应用，以一个游戏实例开发的过程作为指引，分解整个游戏项目开发的框架和详细流程。

本书适合 Android 游戏开发的初学者阅读，也适合作为 Android 游戏开发培训教材和高校相关专业教材或参考书。

因我们水平有限，书中难免有错漏之处，敬请广大读者批评指正。可联系 Email：dd. xy@ 163. com。

编　者

2017 年 4 月

目　录

第一篇　Android 入门

第二篇　Android 游戏开发基础

第三篇 Android 游戏开发应用

第一篇　Android 入门

Android 是什么？它是如何在短短几年内成长为移动市场的最大智能平台的？作为一个开发者，我们先来了解 Android 的前世今生以及今后的前景（第 1 章），然后在第 2 章完成进行 Android 开发的准备工作，并在第 3 章学习创建一个简单而又熟悉的 HelloWorld 项目。

1 Android 简介

读者一定已经听说过 Android，这个名字非常响亮，即使是与软件开发毫不沾边的普通人也听说过这个名字。对很多人而言，Android 意味着一种手机，但更准确地说，Android 是一个包括了操作系统、中间件、用户界面和应用程序的软件平台，是一个为互联网移动终端打造的开放和完整的移动平台。

1.1 Android 的前世今生

Android 的创始人 Andy Rubin，一个对机器人有狂热爱好的天才，有着极其超前的思维和灵感，但当时可能他自己也未必能料到 Android 在短短几年给这个世界带来如此重大的改变。

Andy Rubin 创立了 Danger 和 Android 两个移动操作系统。Danger 以 5 亿美元卖给了微软，成为失败的 Kin 系列；而 Google 公司的收购却让 Android 成为最大的移动操作系统。Google 的野心和眼光让它在 Android 诞生 22 个月后就将其收购，经过两年的秘密开发，2007 年推出 Android 的第一个开放版本。同年，iPhone 一代上市，随后几年 Android 系统版本不断升级，绿色小机器人也慢慢成为与苹果同等地位的标志。短短几年，原来称霸智能手机领域的巨人 Nokia 一蹶不振，微软的 Windows mobile 逐渐消失，高高在上的苹果也只能把市场占有率第一的宝座拱手相让。

Android 的成功在于它的免费开源，Google 的合作厂商可以随意在自己手机上免费搭载 Android 平台，也可以根据自己的需要进行二次开发，这使得众多业界的公司投入了 Android 阵营。

如今的智能移动领域，iOS 和 Android 正打得不可开交，旁观的 Windows mobile 还在摇摆不定。站在如日中天的苹果对面，是由 Google 领衔的一大票业界科技公司：摩托罗拉、三星、联想。相似的场景让人们想起当年苹果与微软和 Intel 的对决，当时与微软同一阵营的公司众多：Dell、HP、宏碁等等。

公司的文化决定着公司的发展路线和决策。注重创新的苹果公司却有着致命的弱点：封闭。用户只能通过苹果公司开放的接口在 iOS 系统上开发第三方应用，但系统本身的开发并不是开放的。苹果公司的理念是可以凭借天才的想法创造最佳的用户体验，然而封闭无法一直走在同行的前面，乔布斯的去世，让很多人对苹果是否能继续创新产生怀疑。

与iOS的封闭相比，Android则走了另一条相反的路线。这是一种建立在Linux上的完全开放和完全免费的平台。全世界优秀的工程师都可以在这之上架设实现自己的想法，这使得整个Android的发展具有无限的可能。几乎所有的手机厂商都推出Android平台的手机，或原生或二次开发，并且在可见的未来还会有更多的公司加入进来。

Andy Rubin给Android命名时，就已经表露了自己对于这个平台的定义：智能机器人。它不但把手机带入智能的领域，未来也会把诸如电视、冰箱乃至家具带入智能领域。

1.2 Android的系统架构

Android平台是基于Linux开发的，整个平台由操作系统、中间件、用户界面和应用软件组成，整个架构采用软件叠层的方式，分成三层，包含五个部分，分别是Linux内核、Android Runtime、Library、Application Framework和Application。

底层以Linux内核为基础，由C语言开发，实现安全性、内存管理、进程管理、网络协议栈和驱动模型等系统核心服务，Android系统在该层还包括内核的各类驱动程序。

中间层包括系统函数库(library)，这部分用C/C ++ 编写，还有一部分用Java语言编写的运行库(runtime)，这部分包括核心库和虚拟机两部分。

最上层也可以细分为两层：应用程序框架(application framework)和应用程序(application)。

应用程序框架就是一般开发者进行Android应用开发的基础，这层提供了很多核心的组件，开发者可以直接使用组件来进行应用程序的快速开发，也可以通过继承对其进行扩展。各种应用软件，包括通话程序、短信程序等组成应用程序层，应用程序由各公司自行开发，以Java语言编写，开发者开发的应用程序均属于这一层。本书也将阐述基于该层进行游戏开发的过程，如图1-1所示。

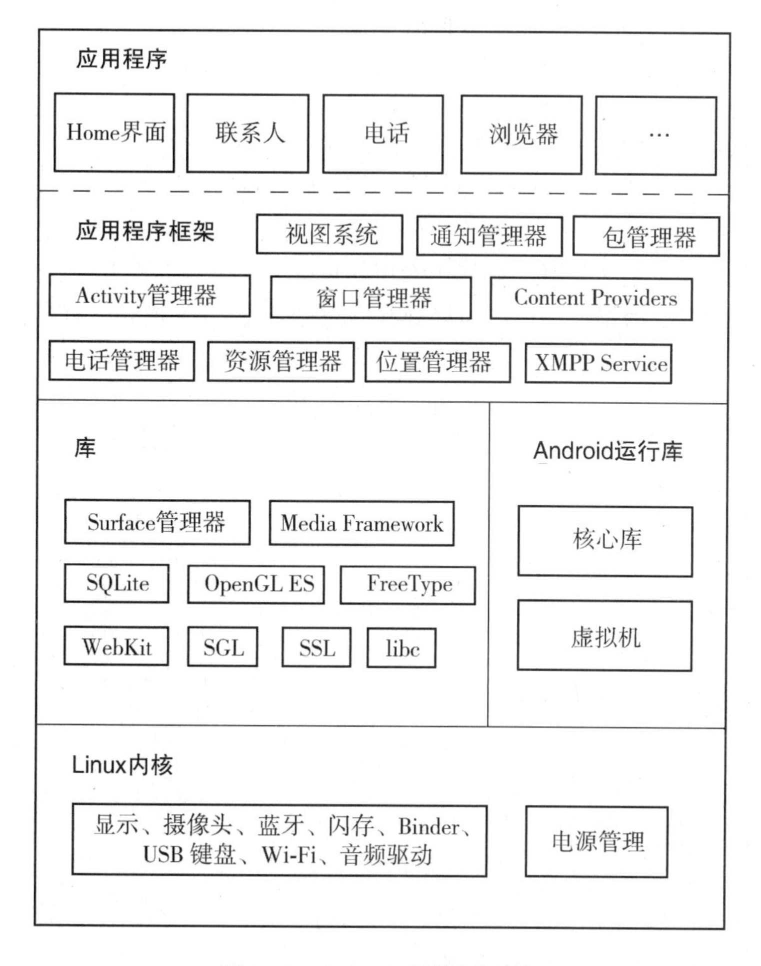

图 1－1 Android 系统层次结构

1.3 Android 游戏开发与外包

游戏是 Android 平台上众多应用类型中上线数量最多的。游戏产品开发的主要实施模式是：①自主开发；②外包开发；③自主＋外包混合开发模式。

随着行业的发展与成熟，为了更快满足玩家和市场需求，开发者必须缩短开发周期，更有效率地利用开发资源，因此外包应运而生。全部或部分采用服务外包模式是企业（发包方）的一种经营战略，是企业（发包方）在内部资源有限的情况下，为取得更大的竞争优势，仅保留最具竞争优势的功能，而其他功能则借助于资源整合，利用外部（接包方）最优秀的资源予以实现。服务外包使企业（发包方）内部最具竞争力的资源和外部（接包方）最优秀的资源结合，产生巨大的协同效应，最大限度地发挥企业自有资源的效率，使企业获得竞争优势并提高对环境变化的适应能力。

简而言之，外包就是做自己最擅长的工作，将不擅长做的工作（尤其是非核心业务）剥离，交给更专业的组织去完成。

作为接包方的企业，若其任务是根据客户需求去开发软件功能，并能够在软件正常运行后提供常规维护和功能扩展开发，而且在开发过程中的需求获取、需求分析、设计、编程、软件测试、版本控制等开发及管理全过程遵循软件工程开发规范，则该类型业务属于信息技术外包（ITO）下的软件设计外包。若任务主要是动漫和游戏的设计（包括创作及制作服务），艺术性强，则属于知识外包（KPO）下的动漫及网游设计研发外包。

希望读者在掌握一种软件开发工具的同时还熟悉软件开发模式，有关服务外包的描述请参阅附录。

2 Android 开发环境

本书介绍的是在 Windows 系统下进行的 Android 开发，相应的工具和环境基本上和 Linux、Mac 系统无太大的区别。

2.1 Android Studio

Google 公司于 2013 年针对 Android 开发推出全新的开发工具 Android Studio。可以登录 Google 网站进行下载（https：// developer. android. com/sdk/installing/studio. html # download）。点击下载 Android Studio V1. 5. 1 for Windows。本书基于目前较新的 1. 5 版，同意其条款后即可下载。如果网络连接不上，可以去国内网站搜索下载。

Android Studio 基于 IDEA 继承环境，提供了集成的 Android 开发工具用于开发调试。安装运行需要 JDK 的支持，进行 Java 语言编程，还需要下载 Java JDK，去 Oracle 公司的页面下载：http：//www. oracle. com/technetwork/java/javase/downloads/index. html。选择 JDK 的下载链接，进行 JDK 版本的选择，我们同样下载最新版本，在点击接受许可协议之后，在众平台中选中开发者相应的系统平台，例如，在 32 位 Windows 下面选择 jdk－7u21－windows－i586. exe 下载并安装。

安装 JDK 后就可以进行 Android Studio 的安装。安装成功后第一次运行时系统会询问是否要导入之前 Android Studio 版本的项目，由于第一次安装，因此没有之前项目可以导入。Android Studio 完成安装界面如图 2－1 所示。

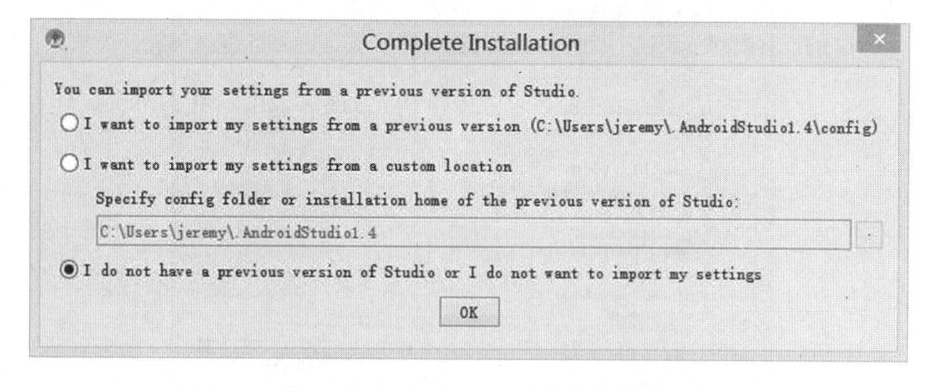

图 2－1　Android Studio 完成安装界面

在学习使用 Android Studio 之前，先看看 Android Studio 安装完成后的目录。该目录提供了各类用于 Android 开发的工具命令以及开发库，如图 2－2 所示。

将选定项目与网络上的其他用户共享。

名称	修改日期	类型	大小
bin	2013/6/6 10:29	文件夹	
config	2013/6/6 11:00	文件夹	
lib	2013/6/6 10:29	文件夹	
license	2013/6/6 10:29	文件夹	
plugins	2013/6/6 10:29	文件夹	
sdk	2013/6/6 10:32	文件夹	
system	2013/6/6 14:40	文件夹	
build	2013/6/6 10:48	文本文档	5 KB
build	2013/5/24 15:16	文本文档	1 KB
build-log	2013/6/6 10:48	HTML 文档	1 KB
LICENSE	2013/5/24 15:16	文本文档	12 KB
studio	2013/5/24 15:16	应用程序	1,416 KB
uninstall	2013/6/6 10:30	应用程序	70 KB

图 2－2　Android Studio 的安装目录

2.2　Android SDK

在 Android Studio 安装目录下有一个 sdk 目录，存放着 Android SDK(Android 的开发库)。Android 发展比较快，它的开发库的版本也比较多。进入 sdk/tools 目录之后执行 android. bat 批处理文件，会直接打开一个 Android SDK Manager 窗口，窗口中列出 Android 开发库的各个版本和工具，我们在列表要下载的项上打勾即可。由于 Android 的升级版本比较多，全部的 SDK 比较大，建议开发者只需选择几个自己开发的目标版本，如图 2－3 所示。

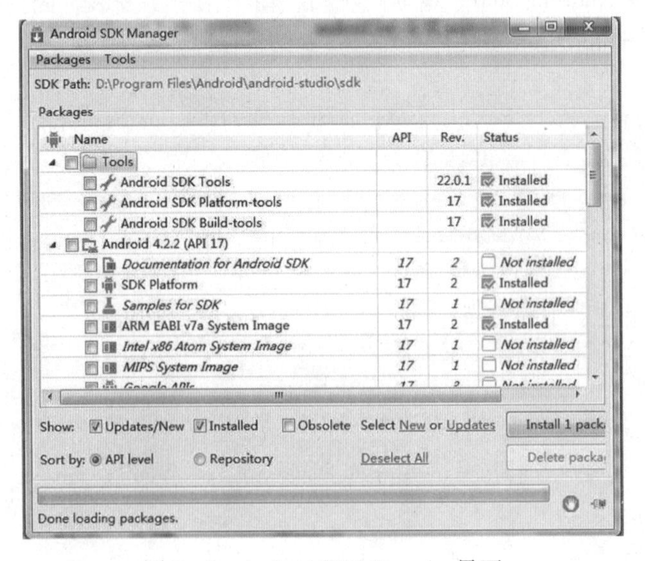

图 2－3　Android SDK Manager 界面

点击“Install 6 packages”按钮，此处6为选中的项数。在弹出的窗口中选择接受许可协议进行安装。如果选择包数比较多，时间会比较长，需要耐心等待。下载开发库之后就可以进入 Android Studio，学习如何创建第一个 Android 项目。

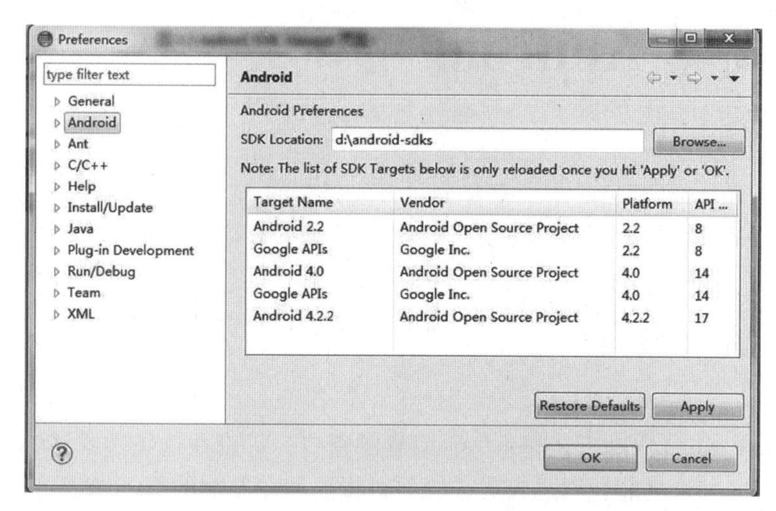

图 2－4　Preferences 环境配置界面

点击左侧列表的 Android 选项(如果没有该选项，说明前面的 ADT 未安装成功)，点击右侧的“Browse”选择已下载好的 SDK 的目录路径，选择正确后，下方列表会刷新当前下载的 SDK 的版本，点击“OK”完成配置。

3 简单 Android 项目

开发一个 Android 项目对于一个有 Java 开发经验的开发者而言，更容易上手。本章介绍如何在 Android 平台上创建并运行经典的 HelloWorld 应用程序。

3.1 创建第一个项目

在 Android Studio 启动后，显示一个引导的窗口，可以在这里选择新建项目或者导入已有项目等，如果之前有创建成功过的项目将显示在左侧的 Recent Projects 中，如图 3－1 所示。

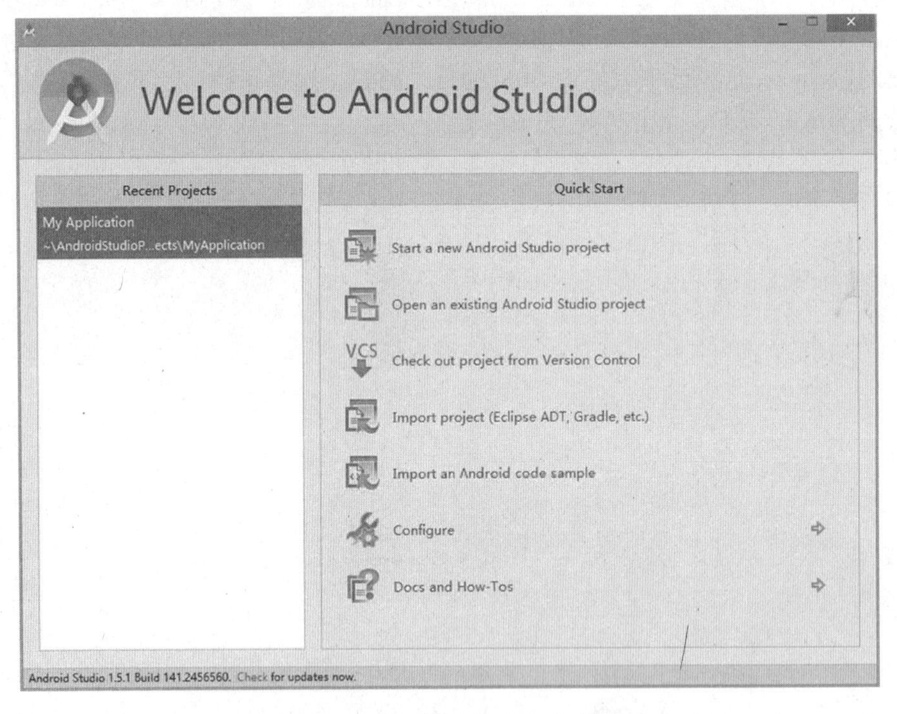

图 3－1　Android Studio 引导页面

选择"Start a new Android Studio Project"创建一个新项目，打开"Create New Project"窗口，如图 3－2 所示。

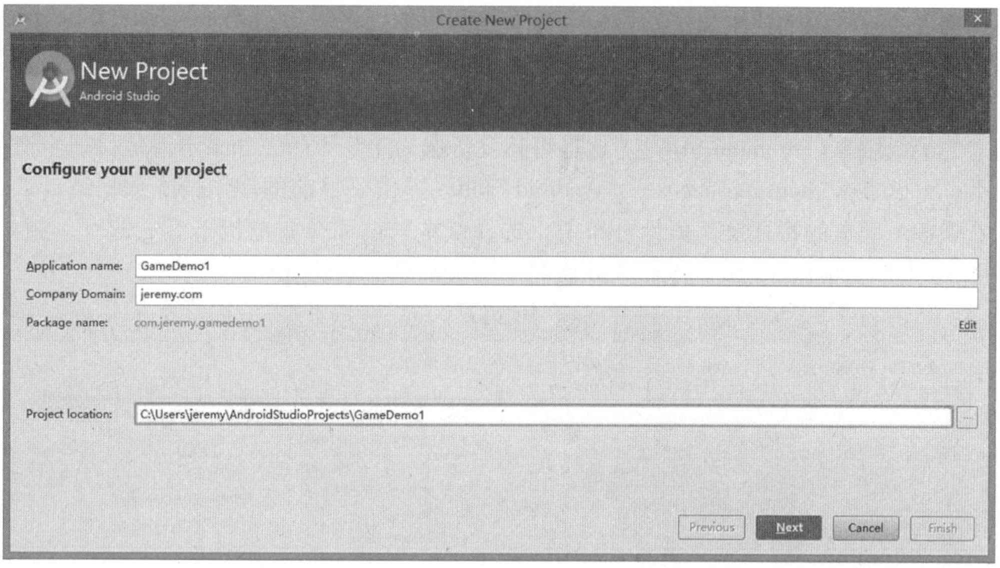

图 3－2　Create New Project 界面

在窗口要设置信息的栏目中，Application name 就是本应用程序发布显示在设备屏幕上的应用名称，也是 Project 的名称。Company Domain 是开发公司的域名。默认情况下 Package name 会将公司域名反过来再加上应用程序名作为包的名称。Package name 就是所在包的名称，可以采用默认的，也可以根据需要通过右侧的“Edit”来指定修改。Project location 是项目在硬盘上的存放路径，可以使用默认路径或自行指定。设置好这些选项后选择“Next”进入下一个页面，如图 3－3 所示。

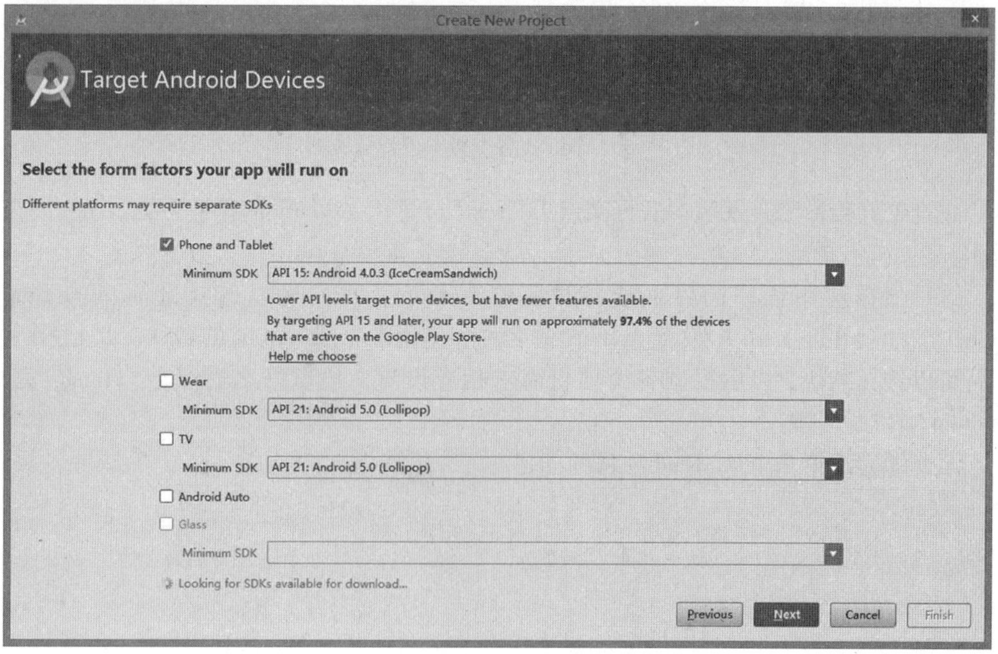

图 3－3　设置后的 Create New Project 界面

该页面用于设置应用程序的目标运行平台及版本。这里包括 Phone and Tablet(手机和平板)、Wear(穿戴设备)、TV(电视)、Android Auto(车载设备)、Glass(Google 眼镜)。我们只需要勾选应用程序打算发行的目标平台即可。每种设备都可以指定该应用需求的最低版本(Minimum SDK)。这里将其设备选为 4.0.3 版，如果不清楚目前主流的版本，可以点击“help me choose”。Android Studio 提供了目前的使用情况数据统计，可以看到各个版本的特征及其累计分布值，选择版本较高而分布值也不低的即可。例如，在图 3-4 中，根据显示的内容，4.0 版和 4.1 版都是适当的选择。

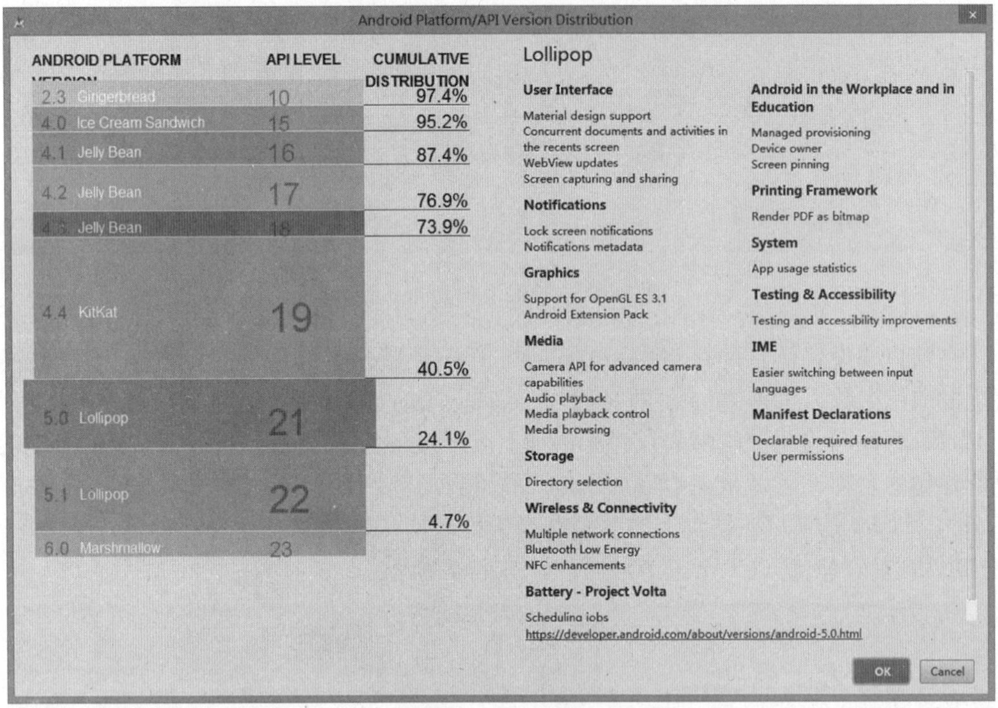

Android Platform/API Version Distribution

ANDROID PLATFORM VERSION	API LEVEL	CUMULATIVE DISTRIBUTION
2.3 Gingerbread	10	97.4%
4.0 Ice Cream Sandwich	15	95.2%
4.1 Jelly Bean	16	87.4%
4.2 Jelly Bean	17	76.9%
4.3 Jelly Bean	18	73.9%
4.4 KitKat	19	40.5%
5.0 Lollipop	21	24.1%
5.1 Lollipop	22	4.7%
6.0 Marshmallow	23	

Lollipop

User Interface
Material design support
Concurrent documents and activities in the recents screen
WebView updates
Screen capturing and sharing

Notifications
Lock screen notifications
Notifications metadata

Graphics
Support for OpenGL ES 3.1
Android Extension Pack

Media
Camera API for advanced camera capabilities
Audio playback
Media playback control
Media browsing

Storage
Directory selection

Wireless & Connectivity
Multiple network connections
Bluetooth Low Energy
NFC enhancements

Battery - Project Volta
Scheduling jobs
https://developer.android.com/about/versions/android-5.0.html

Android in the Workplace and in Education
Managed provisioning
Device owner
Screen pinning

Printing Framework
Render PDF as bitmap

System
App usage statistics

Testing & Accessibility
Testing and accessibility improvements

IME
Easier switching between input languages

Manifest Declarations
Declarable required features
User permissions

OK　Cancel

图 3-4　Android 各个版本累计分布图

在设置好目标平台及版本后再次点击“Next”，进入 Activity 的创建页面，如图 3-5 所示。

这里根据应用程序的界面选择相应的 Activity，我们可以选择最简单的 Empty Activity，而后在图 3-6 中设置 Activity 的名称和使用的布局 Layout 的名称，此处第一个项目可以选择默认，以后根据各个项目需求相应设置。设置好之后点击“Finish”完成 Android 项目的创建。

至此即完成第一个 Android 项目的创建。

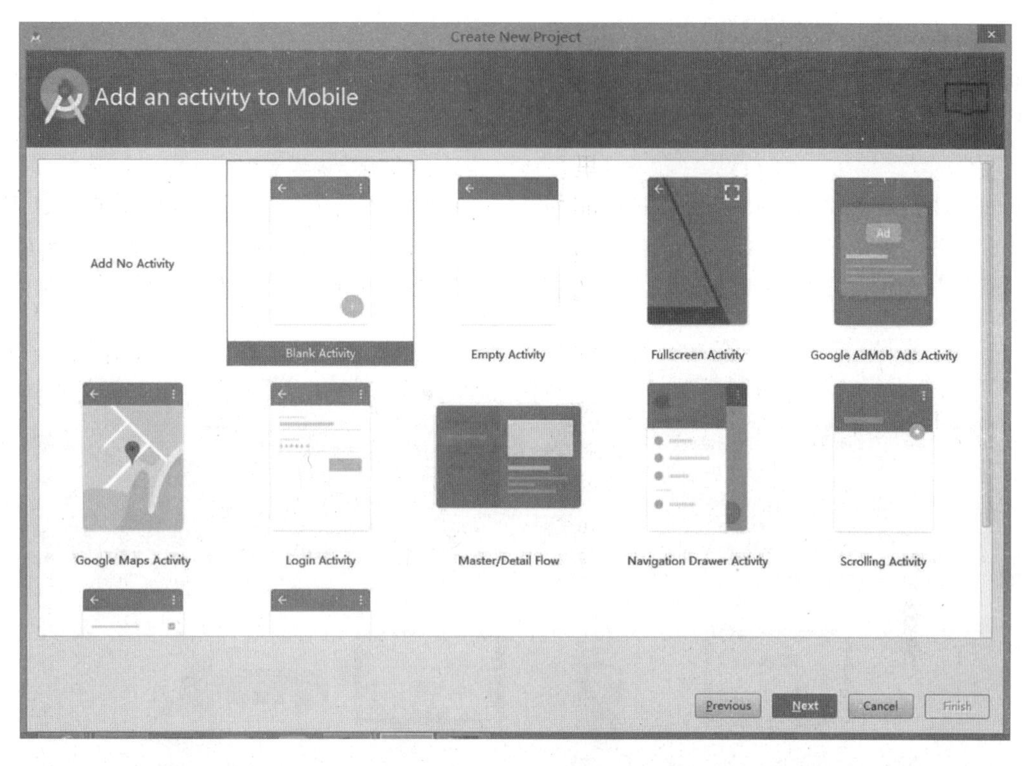

图3－5　Activity 的选择

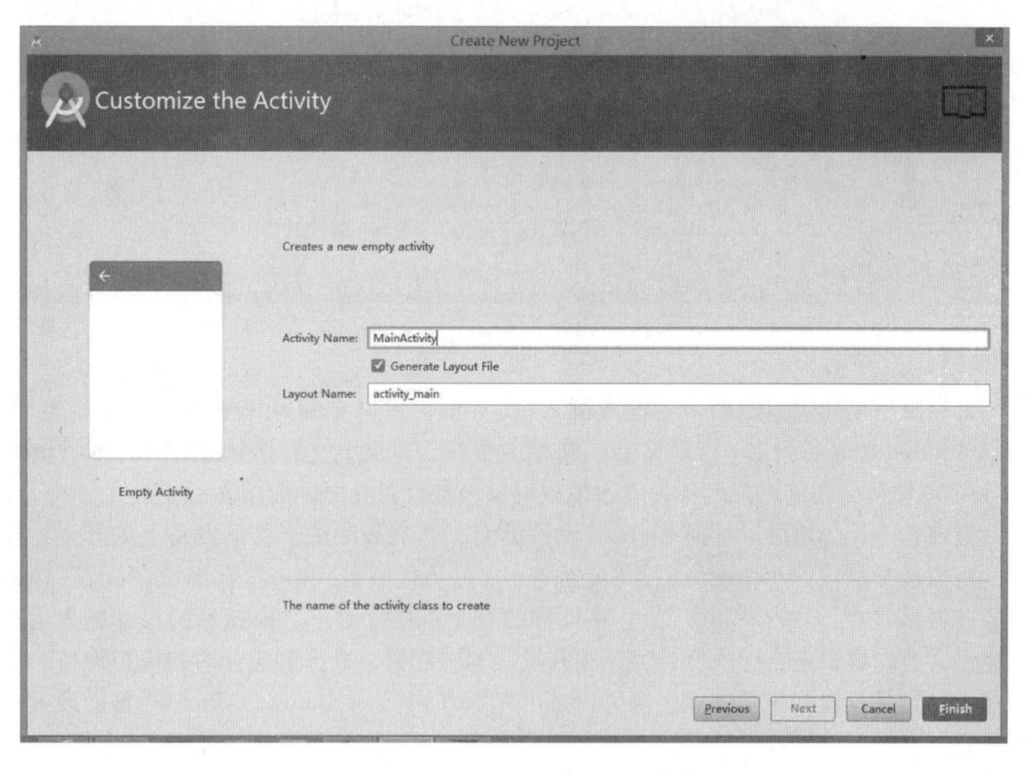

图3－6　设置 Activity 和 Layout 名称

3.2 建立 Android 模拟器

想让创建的项目运行，需要通过 Android 设备或者虚拟设备，在前期的简单应用开发学习中，可以使用虚拟设备——模拟器来运行项目，选择菜单"Tools"中的"Android""AVD Manager"菜单项，打开模拟器管理窗口，在该窗口中创建和管理虚拟设备，如图 3-7 所示。

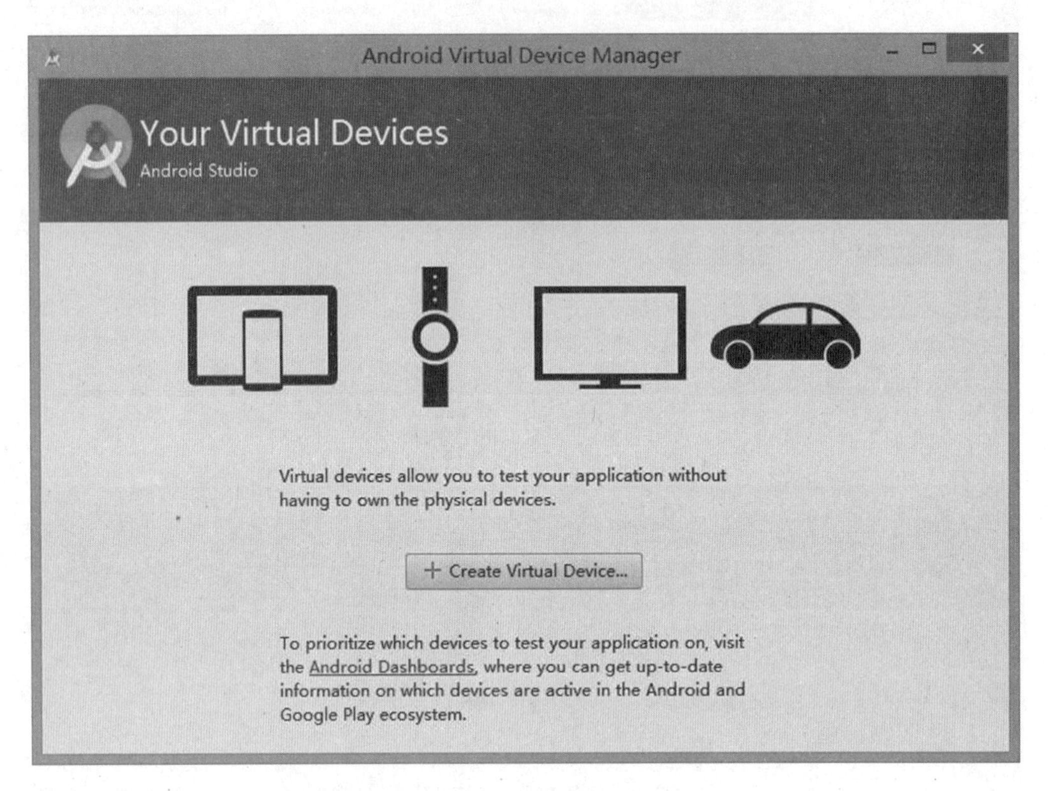

图 3-7　Android 模拟器管理界面

如果没有创建过模拟器，会进入图 3-7 界面，可以看到可供创建的模拟器类型，包括 Android 的各种设备(移动设备、穿戴、电视、车载)，点击中间的"Create Virtual Device"创建一个新的模拟设备，进入模拟设备参数设置界面，如图 3-8 所示。

在窗口左侧仍然可以切换 Android 设备类型，本书只专注于移动设备上的开发。在中间可以选择默认的几种硬件设备(包括 Nexus 的历代版本)和一些标准设备配置(包括设备的屏幕尺寸、分辨率和密度)。在同样的开发机器配置下，模拟器的尺寸越大运行就越慢，所以要根据自己的发行目标和开发机器的配置，选择创建合适的模拟设备。如果提供的默认设置列表不满意，也可以点击左下方的"New Hardware Profile"自行设置这些参数。在选择好屏幕参数后，点击"Next"，进入下一步参数配置，如图 3-9 所示。

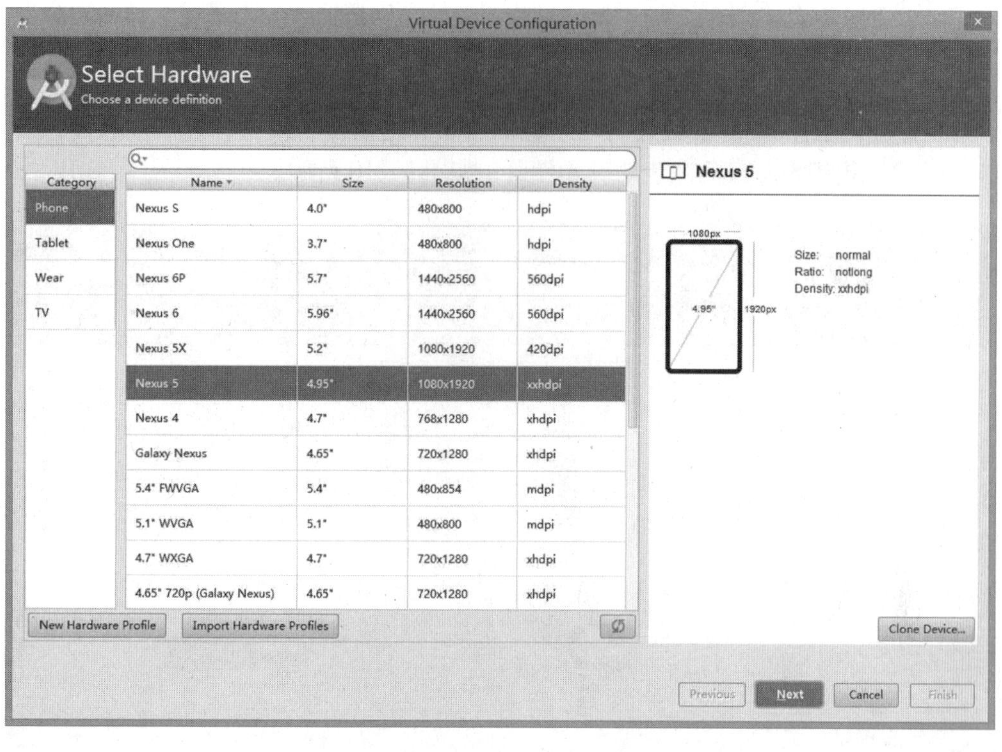

图 3－8　模拟器参数设置界面

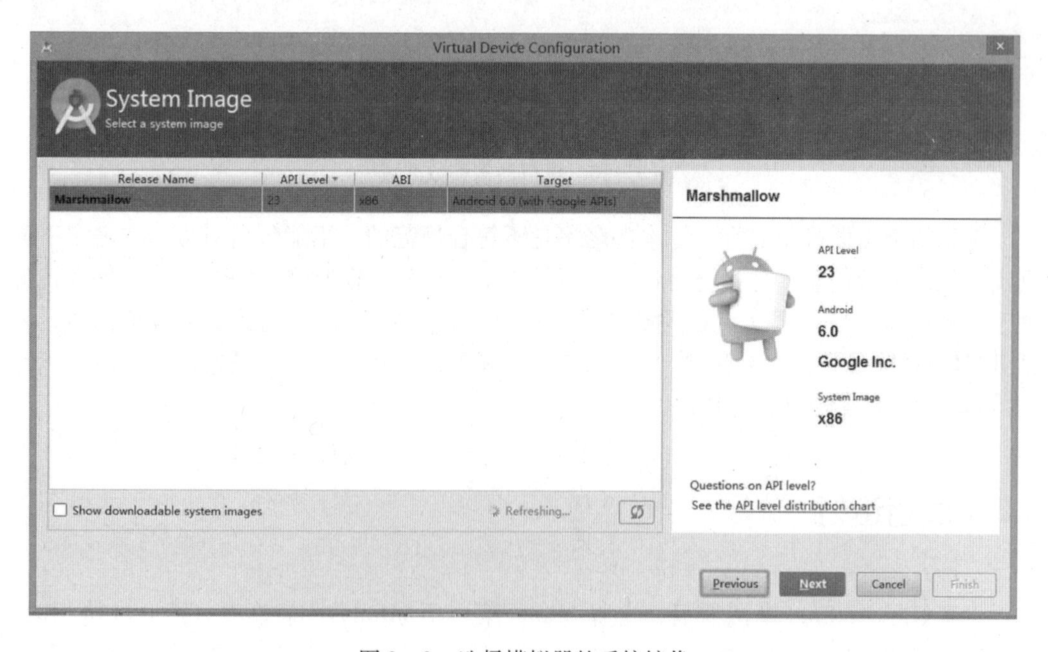

图 3－9　选择模拟器的系统镜像

在该界面中，需要选择该模拟设备的运行 Android 版本、系统镜像，这里出现的版本和镜像与在 SDK 中安装的版本相关，在图 3－9 中只安装了 SDK6.0 的版本和相应的 x86 的镜像，所以只能选择 6.0 的 x86 镜像。由于电脑是使用 x86 的处理器，因此选择 x86 的镜像会比选择 ARM 的镜像的启动速度和运行速度更快。选择好版本和镜像后点击“Next”进入模拟设备最后一步的设置界面，如图 3－10 所示。

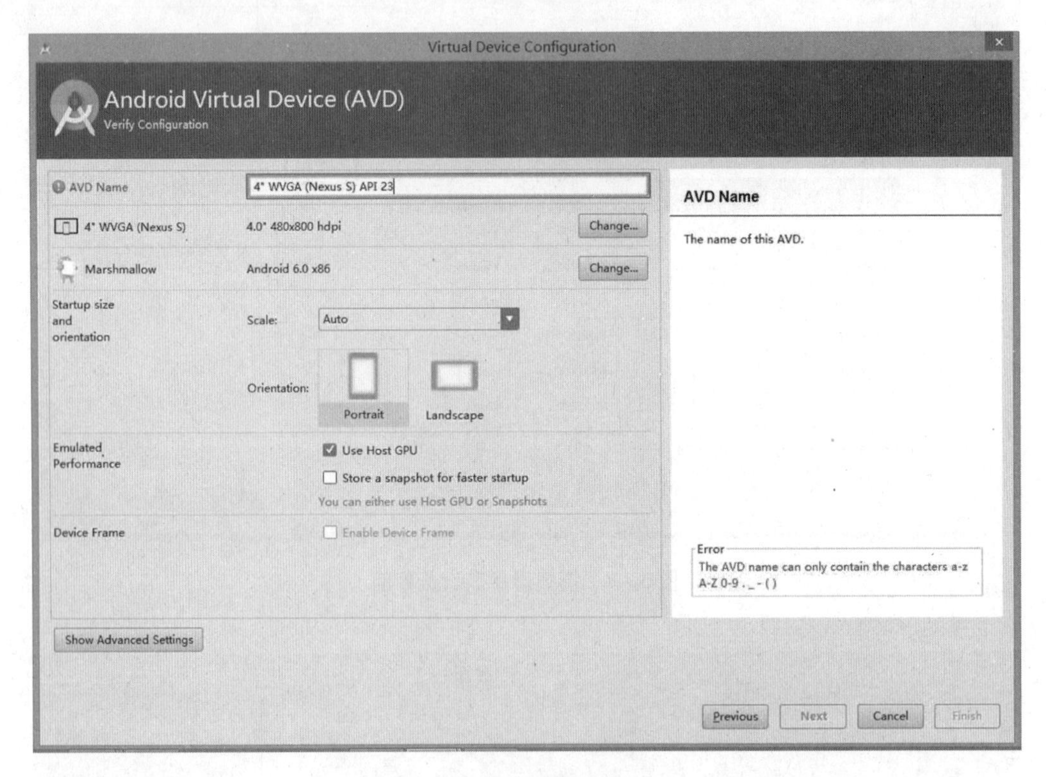

图 3－10　模拟器的参数配置

首先给模拟器换个名字，原本的名字是不符合规范的，不能包括引号。在名字下面列出其他的参数信息，包括屏幕信息、运行版本、缩放、横竖屏、GPU 加速、保存镜像用于启动加速，甚至可以点击下面的“Show Advanced Settings”，看到更多设置信息，包括 RAM、系统缓存、内部存储、外部 SD 卡等设置。在初学阶段，设置好名字后可以点击“Finish”来完成模拟器的创建过程。

3.3　项目运行

设置好了 Android 模拟器后，选择“Run”菜单下“Run…”菜单项，第一次执行时，会弹出 Run 窗口，点击“Edit configurations”打开运行配置窗口，选择第一个 Android Application，点击“Run”按钮运行，将出现选择模拟器的界面，如图 3－11 所示。

Device Chooser

Choose a running device

Device	State	Compatible	Serial Num...

Nothing to show

Launch emulator

Android virtual device: avd4

Use same device for future launches

OK Cancel Help

图 3－11　运行设备选择界面

凡是支持该应用程序的模拟器均可以在下方的 Launch emulator 中选择。如果有已经运行的模拟器或设备，会在上方的 running device 列表中看到。选择好模拟器后点击“OK”使用该模拟器运行 App，就可以看到系统自动打开模拟器并运行 HelloWorld 项目。模拟器的启动比较慢，所以调试的过程中不要每次测试后都关闭，这样可省去启动的时间。启动后，出现一个询问是否要监测 LogCat 的窗口，选择“Yes”。模拟器运行效果如图 3－12 所示。

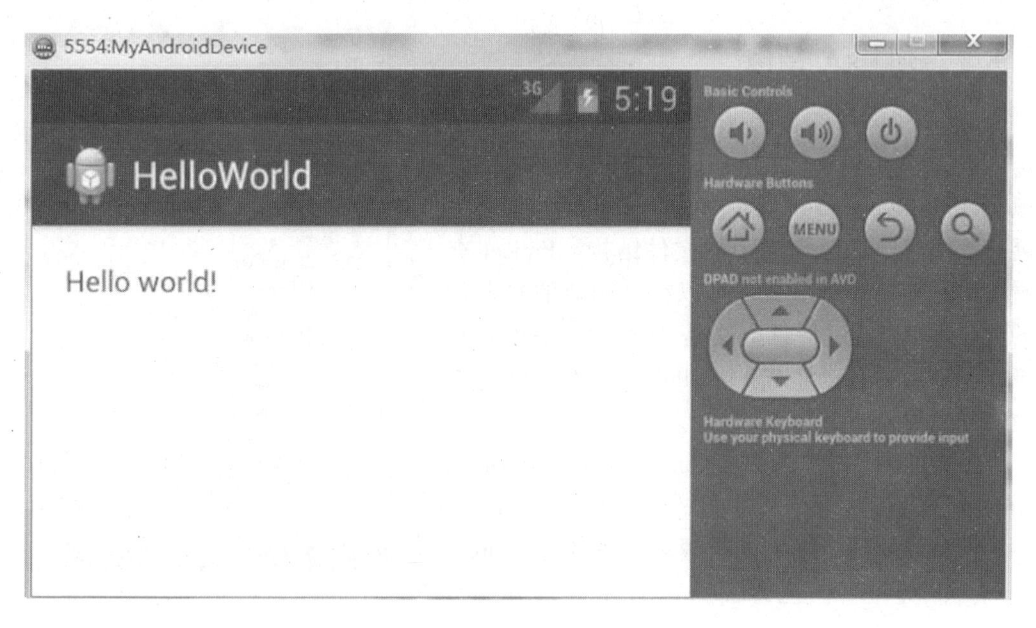

图 3－12　模拟器运行图

模拟器会在计算机上显示一个 Android 手机画面，画面上有一部 Android 手机应用的各类按键，包括 Home、菜单、回退和搜索，以及音量调节和关机键、方向键。为了调试方便，在键盘上也有按键对应这些手机按键的功能，甚至可以通过键盘模拟手机横竖屏的效果，如表 3－1 所示。

表 3－1　Android 模拟器手机键盘按键对应表

手机按键	键盘按键
音量减少	Ctrl + F5/" + "
音量增加	Ctrl + F6/" - "
电源键	F7
Home	Home
Menu	F2/PageUp
回退	Esc
搜索	F5
方向键：上、下、左、右、中	数字键 8、2、4、6、5
旋转屏幕	数字键 7、9
全屏、窗口切换	Alt + Enter

3.4　项目结构解析

前面我们利用程序向导轻松建立了第一个 Android 项目 HelloWorld。本节我们通过解析该项目的结构，来了解 Android 项目的基本组成部分，以及各部分的功能。

3.4.1　项目文件结构

Android 项目文件结构和其他的 Java 项目一样，以树形结构的方式组织。项目创建成功后可以在 Android Studio 环境左侧的 Project 信息窗口中展开该项目，显示项目中的文件结构。Project 查看窗口有多种查看方式，可以在该窗口的左上角的下拉框中选择，默认方式是 Android。在该方式下，可以看到自动生成的 HelloWorld 中已经包含很多的文件，如图 3－13 所示。

图 3－13　Project 窗口

最大的两个分类分别是 App 和 Gradle Scripts。App 就是当前这个项目中的一个模块(module)。一个项目(project)中可以有很多个模块。Andorid Studio 和 Eclipse 在这方面略微有不同，Android

Studio 只能打开一个项目。实际上这里的工程相当于 Eclipse 的 workspace，而 Eclipse 中的 project 也可参照 module 理解。在当前这个项目中，App 就是显示 HelloWorld 功能的一个 module。

而 Gradle 则是 Android Studio 提供的构建工具，Gradle 版本更新比较快，因此，安装好 Android Studio 后，在创建第一个项目时，往往需要更新 Gradle，需要去联网下载，下载时 Android Studio 环境下方状态栏处会出现一个进度条。这个过程是自动的，只要连上互联网即可。在这部分中有两个 build. gradle 分别对应整个项目和 App 这个 module 的 gradle 配置信息。gradle-wrapper. properties 声明了 Gradle 的目录与下载路径以及当前项目使用的 Gradle 版本。settings. gradle 是全局的项目配置文件，里面主要声明了一些需要加入 Gradle 的 module。

在此把重点放在 App 这个 module 上。将其展开后，按照从上往下的顺序，首先是 manifests 清单文件，这是一个 XML 文件，里面包括了本项目重要的配置信息，应用程序通过这个文件向 Android 系统提供一些必要的信息，比如包的名称、组件描述已经具备的权限等等。紧随其后的是 java 目录，其中存放的就是项目的源文件，在 HelloWorld 项目中，目前这里只有一个源文件，即 MainActivity. java。

3.4.2 资源文件

源文件目录下面是 res 目录，即资源目录，游戏开发中用到的声音、图片、图标、模型等都属于资源。在 Android 项目中有个资源标记文件 R. java，其中存放着 Android 项目用到的各类资源的标记，该文件在 project files 方式下的项目结构中可见，Android 方式下看不到。当然这个文件是由 Android 项目自动维护和生成的，开发时不用去修改它。每导入一个新的资源，会在 R. java 中自动生成一行新的代码，定义一个新的以资源文件名命名的常量来标记该资源，以便在项目中调用。res 目录下的 drawable 目录包括图像或图标文件，Android 会自动为 drawable 目录下的图片文件生成 ID，例如，在该目录下放入一个 sample. png，那么 R. java 会自动生成一行代码：

```
public static final int sample = 0x7f020001;
```

导入项目中的资源文件只能以小写字母和下画线作首字母，随后的名字中只能出现 [a～z 0～9_.] 这些字符，否则导入会失败。

在项目中可以看到 res 目录下，drawable 表示可绘制资源，包括图片和图标，layout 表示布局，value 表示定义的各类值。而 mipmap 也是存放可绘制资源，与 drawable 的不同之处在于它针对不同的 dpi 有不同像素尺寸的图片，在渲染需要缩放时会有更高的效率，项目自动将启动图标放在这里了。项目存在多个 mipmap 系列目录，其名称后缀分别是 hdpi、mdpi、xhdpi、xxhdpi、xxxhdpi。dpi 指每英寸[*]的像素个数，它直接代表屏幕的显示精度。Android 作为一个开放的系统，旗下设备的硬件参数五花八门，从 2.8 英寸的手机到 46 英寸的 Google TV 设备各不相同。对于一个宽高固定的 UI 组件而言，在显示精度较低的屏幕上显得很大，而在显示精度相对较高的屏幕上则显得很小。对于图

* 英寸为非法定计量单位，1 英寸≈2. 54 厘米。

片而言，系统对不同的分辨率会自动对图片进行缩放以适配具体的屏幕分辨率。但是考虑到用户的体验，开发者应该考虑其应用在各种屏幕上的显示效果，在图片资源上也就产生了相应的分级目录。不同分级目录中的图片对应不同的显示精度，开发应用程序时，开发者提供不同尺寸的图片对应不同目录，最早的版本只有 ldpi、mdpi、hdpi，分别代表低、中、高分辨率。随着硬件的飞速提升，逐渐有了 xhdpi 超高分辨率。本书采用的最新 SDK 版本已有 xxxhdpi 级别，对应4k 高清的分辨率，而相应的低分辨率的 ldpi 已取消。

3.5 项目源码

了解 Android 项目的文件组织结构后，再来学习其中比较重要的一些代码文件。目前项目中只有一个 java 源文件，就是在创建项目时设置过名称的 MainActivity. java。另外有个清单文件 AndroidManifest. xml 和布局文件 activity_ main. xml 文件。

3.5.1 AndroidManifest. xml

AndroidManifest. xml 文件是应用程序的描述文件，包括整个应用程序提交给运行系统的信息。代码如下：

```
<?xml version = "1.0" encoding = "utf - 8"? >
<manifest xmlns:android = "http://schemas.android.com/apk/res/android"
   package = "com.jeremy.helloworld" >
   <application
       android:allowBackup = "true"
       android:icon = "@ mipmap/ic_launcher"
       android:label = "@ string/app_name"
       android:supportsRtl = "true"
       android:theme = "@ style/AppTheme" >
       <activity android:name = ".MainActivity" >
            <intent - filter >
                 <action android:name = "android.intent.action.
                  MAIN"/ >
                 <category android:name = "android.intent.category.
                  LAUNCHER"/ >
            </intent - filter >
       </activity >
   </application >
</manifest >
```

作为一个 .xml 文件，标签代表不同的含义，而标签的属性代表相应的各类配置信息。首先是 manifest，其中设置了 Android 的命名空间和包的名称。Application 代表应用程序，该标签声明了应用程序的组件及其属性。上面的例子中设置了允许系统备份，应用程序图标以及指定应用程序名、主题和布局方式。其中包含了 activity(活动)标签，activity 是应用程序的组件并设置了自身的一些信息(在下一章会详细介绍)。当前项目只包含了这一个组件。

3.5.2 MainActivity.java

目前的项目中只有简单的一些 java 源码，均在该文件中，内容如下：

```
package com.jeremy.helloworld;

import android.support.v7.app.AppCompatActivity;
import android.os.Bundle;

public class MainActivity extends AppCompatActivity {

    @Override
    protected void onCreate(Bundle savedInstanceState) {
        super.onCreate(savedInstanceState);
        setContentView(R.layout.activity_main);
    }
}
```

可以看到源码中定义了一个 MainActivity 类，继承自 Android 库提供的 AppCompatActivity。这个类就是当前这个程序的主活动类，事实上也是目前唯一的活动类。在该类中重写了一个方法 onCreate，其中代码是在活动对象创建时要给该 Activity 加载指定的内容视图，就是下一节的 activity_main.xml 文件代表的布局。

3.5.3 activity_main.xml

activity_main.xml 文件在工程 res/layout 中，代表着一个布局资源。内容如下：

```
<?xml version="1.0" encoding="utf-8"?>
<RelativeLayout
    xmlns:android="http://schemas.android.com/apk/res/android"
    xmlns:tools="http://schemas.android.com/tools"
    android:layout_width="match_parent"
    android:layout_height="match_parent"
    android:paddingLeft="@dimen/activity_horizontal_margin"
    android:paddingRight="@dimen/activity_horizontal_margin"
    android:paddingTop="@dimen/activity_vertical_margin"
```

```
        android:paddingBottom="@dimen/activity_vertical_margin"
        tools:context="com.jeremy.helloworld.MainActivity">

        <TextView
            android:text="Hello World!"
            android:layout_width="wrap_content"
            android:layout_height="wrap_content" />
</RelativeLayout>
```

在 Android Studio 开发环境中，直接双击 activity_main.xml 打开其 Design 视图，切换到 Text 视图就可以看到上述代码。这也是一个 .xml 文件，RelativeLayout 标签指定了一种布局方式——相对布局，而其中的 TextView 则是该布局文件中唯一的 UI 元素，该应用程序使用这个文本框控件显示 HelloWorld 文字。

第二篇　Android 游戏开发基础

在学习了 Android Studio 开发环境和成功创建第一个简单 Android 项目之后，即可进行 Android 游戏开发。本篇首先阐述 Android 开发的基础知识，然后针对游戏开发各个模块进行详细介绍。

4 Android 组件

在开始学习 Android 游戏编程之前，应先了解 Android 开发中的一些基础知识。Android 中常用的组件有 Activity、Service、Broadcast Receiver、Content Provider，负责组件之间通信的是 Intent。

4.1 Activity

4.1.1 Activity 组件概述

Activity 在组件中是最常用的，只要有界面的应用程序，都会用到 Activity。它代表一个屏幕，可以通过 setContentView 方法装载指定的视图(view)进行显示，完成与用户的交互功能。比较简单的应用开发，可以把应用程序的每一个界面做成不同的 Activity，界面的切换通过 Activity 的跳转来实现。应用中的 Android 组件都需要在 AndroidManifest.xml 中注册才能使用，例如，在 HelloWorld 项目中，首先需要定义一个继承自 AppCompatActivity 的 MainActivity 类，然后在 AndroidManifest.xml 中添加相应的标签，设置好属性完成注册过程。当然，这个过程在项目创建过程中会自动完成。如果需要更多的组件，就需要自行完成这两步才能使用。

4.1.2 Activity 生命周期

Android 是一个多任务的操作系统，Android 手机用户都有过类似的体验，一边听音乐一边看电子书，或者玩游戏时突然来电话。总之，并不需要结束一个应用才能开启下一个，而每一个应用都会占用系统资源，如果没有一个合理的任务管理机制，随着任务开启数量的增多，系统就会被拖慢乃至崩溃。实际上，Android 为组件设立了生命周期。而生命周期并不需要用户也不需要应用开发者管理，而是由系统来负责。但是，应用开发者掌握系统如何管理生命周期是至关重要的，否则，开发的应用在被系统强制销毁之前，应用却未能有及时相应的处理(比如保存数据)，那么给用户带来的体验将是非常糟糕的。

一般情况下，每个应用程序运行都会产生一个进程(process)，系统内存不足，系统会自动对进程进行回收(销毁)。运行在 Android 系统下的进程有以下五种：

- 前端进程。是拥有一个显示在屏幕最前端并与使用者做交互的 Activity(它的 onResume 已被调用)的进程，也可能是一个拥有正在运行的 IntentReceiver(它的

onReceiveIntent( )方法正在运行)的进程。在系统中，这种进程是很少的，只有当内存低到不足以支持这些进程的继续运行，才会将这些进程销毁。通常这时设备已经达到了需要进行内存整理的状态，为了保障用户界面不停止响应，只能销毁这些进程。

- 可视进程。是拥有一个用户在屏幕上可见的，但并没有在前端显示的 Activity(它的 onPause 已被调用)的进程。例如，一个以对话框显示的前端 Activity 在屏幕上显示，而它后面的上一级 Activity 仍然是可见的。这样的进程是非常重要的，一般不会被销毁，除非为了保障所有的前端进程正常运行才会被销毁。
- 服务进程。是拥有一个由 startService( )方法启动的 Service 的进程。尽管这些进程对于使用者是不可见的，但它们做的通常是使用者所关注的事情(如后台 MP3 播放器或后台上传下载数据的网络服务)。因此，除非是为了保障前端进程和可视进程的正常运行，否则系统不会销毁这种进程。
- 后台进程。是拥有一个用户不可见的 Activity(onStop( )方法已经被调用)的进程。这些进程不直接影响用户的体验。如果这些进程正确地完成了自己的生命周期(详情可参考 Activity 类)，系统会为了以上三种类型进程而随时销毁这种进程以释放内存。通常会有很多这样的进程在运行着，因此这些进程会被保存在一个 LRU 列表中，以保证在内存不足时，用户最后看到的进程将在最后才被销毁。
- 空进程。是那些不拥有任何活动的应用组件的进程。保留这些进程的唯一理由是，作为一个缓存，在它所属的应用的组件下一次需要时缩短启动的时间。同样，为了在这些缓存的空进程和底层的核心缓存之间平衡系统资源，系统会经常销毁这些空进程。

从用户的角度而言，进程的感受不如任务(task)的感受更明显。Android 系统的运行，其组件是可以跨不同的应用来调用的。例如，在某个游戏中需要访问某个站点，那么可以在游戏界面中调用系统中已有的浏览器来访问该站点，而不需要游戏开发者在游戏开发过程中自己写一个网页浏览的模块。这种情况很多，当用户执行某个应用的过程时，可能会开启到其他的应用，这样应用之间的调用关系就会变得复杂。为了应对这种情况，Android 引入了 task 概念。在用户运行一个应用时，就启动一个 task，而一个 task 对应一个 Activity 栈。当前应用启动后，其启动 Activity 将进入栈中，当该 Acitivity 调用其他 Activity(无论是否是本应用的 Activity)时，新调用的 Activity 将压入栈，占据栈顶，而原来的 Activity 下沉。当用户按“Back”回退时，当前 Activity 出栈，原来的 Activity 占据栈顶并显示在屏幕上。

Activity 在其生命周期中，有以下四个状态：

- Active(活动)状态。Activity 运行于屏幕最前端，处于栈的最顶端。此时它处于可见并可与用户交互的激活状态，同一时刻只会有一个 Activity 处于该状态。
- Paused(暂停)状态。当 Activity 被另一个透明或者 Dialog 样式的 Activity 覆盖时的状态。它仍然可见，但它已经失去了焦点，不可与用户交互。此时还保留着所有的状态。
- Stopped(停止)状态。当 Activity 被另外一个 Activity 覆盖、失去焦点并不可见时处于 Stopped 状态。此时仍保留着所有的状态。
- Dead(死亡)状态。Activity 被系统杀死回收或者没有被启动时处于 Killed 状态。

对于一个 task 而言，其中调用的 Activity 在出栈入栈的过程中其状态转换如图 4－1 所示。

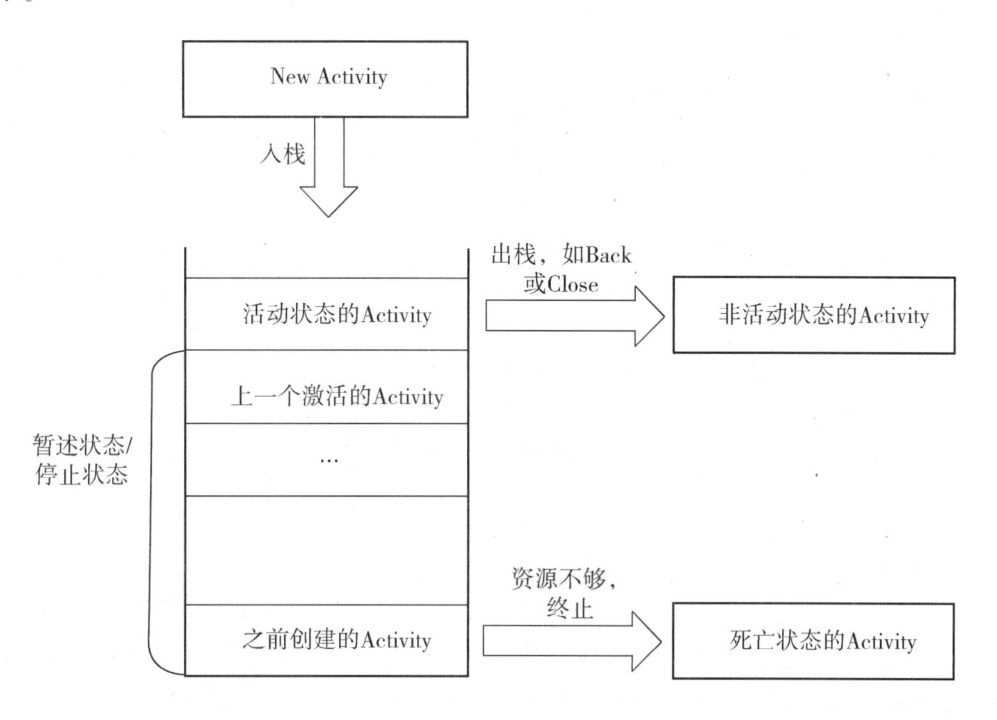

图 4－1　task 中 Activity 的状态变迁

为了让开发者完善地掌握 Activity 的状态变化，在 Activity 的生命周期中，状态发生改变时，系统会自动调用 Activity 的七个方法。调用流程如图 4－2 所示。

从图 4－2 可以看出，七个方法分别在生命周期的不同阶段被调用，可以帮助我们获悉 Activity 的状态变化。而开发者只需要在应用的 Activity 中重写这七个方法，加入处理代码，就可以在 Activity 的状态改变时做出相应的处理。七个方法说明如下：

- void onCreate(Bundle savedInstanceState)，当 Activity 第一次被实例化时系统会调用。一个生命周期只调用一次这个方法。该方法中的代码通常用于初始化设置，例如，在 HelloWorld 项目中，该方法已经重写，在其中装载布局。
- void onStart()，当 Activity 初始化完成时，此时可见但未获得用户焦点不能交互。
- void onRestart()，当 Activity 已经停止然后重新被启动时系统会调用。
- void onResume()，当 Activity 可见且获得用户焦点能交互时系统会调用。
- void onPause()，当 Activity 从活动状态到暂停状态时调用。
- void onStop()，当 Activity 从暂停状态到停止状态时调用。
- void onDestory()，当 Activity 被回收时调用。

为了更好地理解 Activity 生命周期，举例说明：有两个 Activity 分别为 A 和 B，首先启动 A，在过程①，该任务将 A 入栈(在这之前栈空)，A 的方法调用顺序为：

A：onCreate→A：onStart→A：onResume

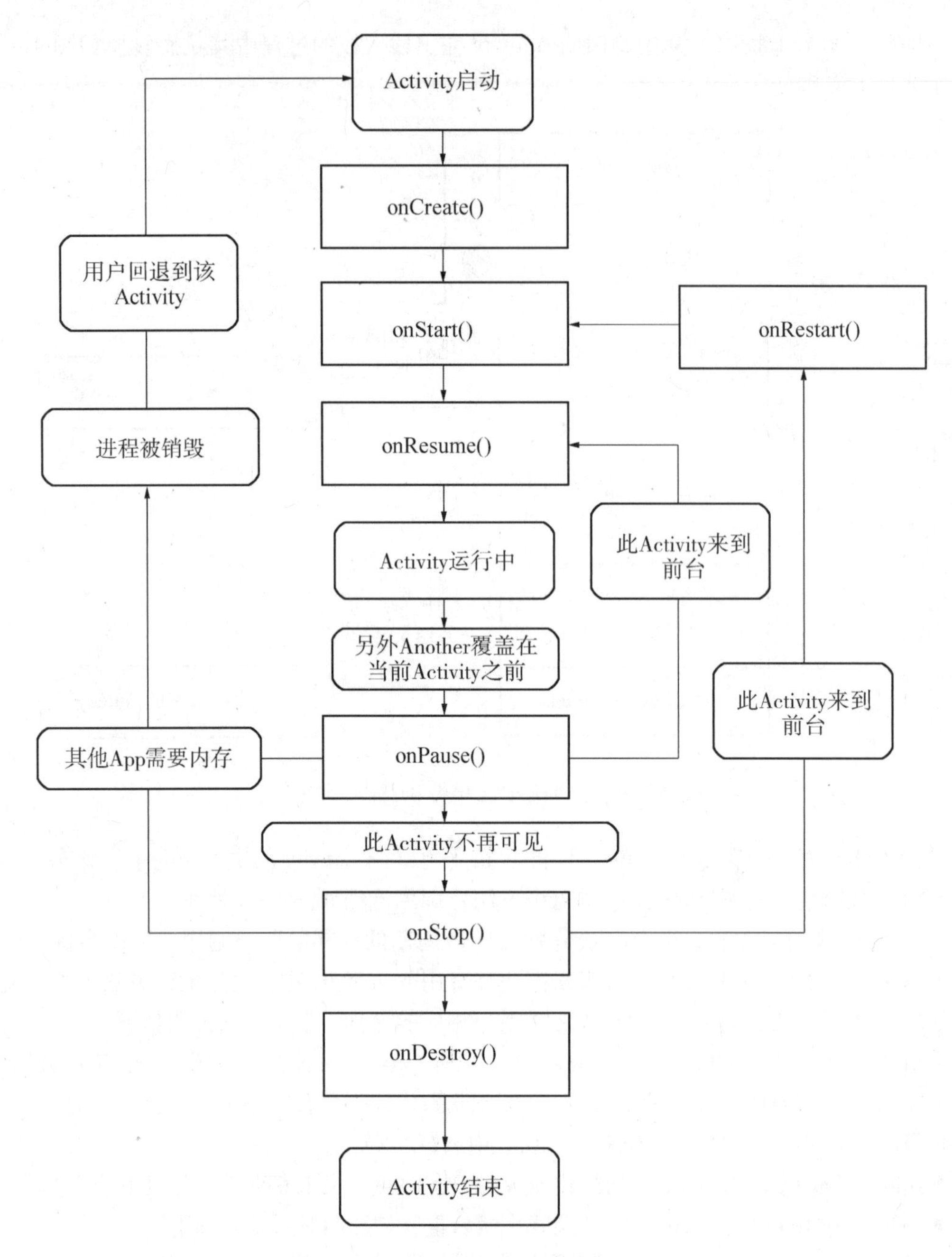

图4－2　Activity 生命周期

在这个过程中，A 的状态从 Dead 转换到了 Active。此时从 A 跳转打开 B，B 将入栈，A 下移，从而 B 在栈顶。过程②调用顺序为：

A：onPause→B：onCreate→B：onStart→B：onResume→A：onStop

在跳转完成后，过程③用“Back”键从 B 回到 A。这时 B 将退栈并回收，A 重新回到栈顶。

B：onPause→A：onRestart→A：onStart→A：onResume→B：onStop→B：onDestory 变化示意图如图 4－3 所示。

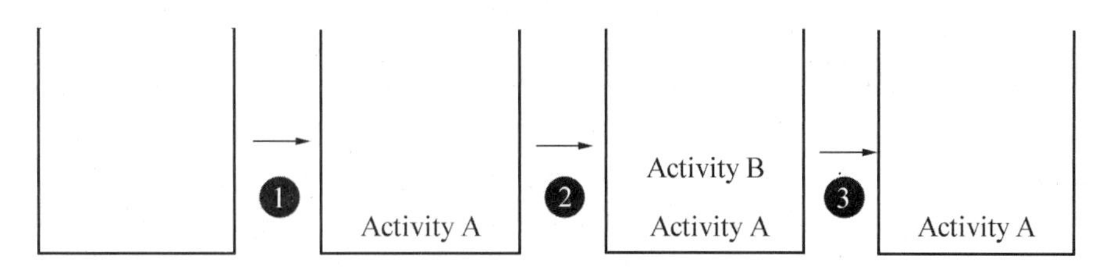

图 4－3　栈变化示意图

4.1.3　LogCat

在软件开发中，日志输出是很常用的调试手段，Android Studio 提供了 LogCat 来输出日志信息，如图 4－4 所示。

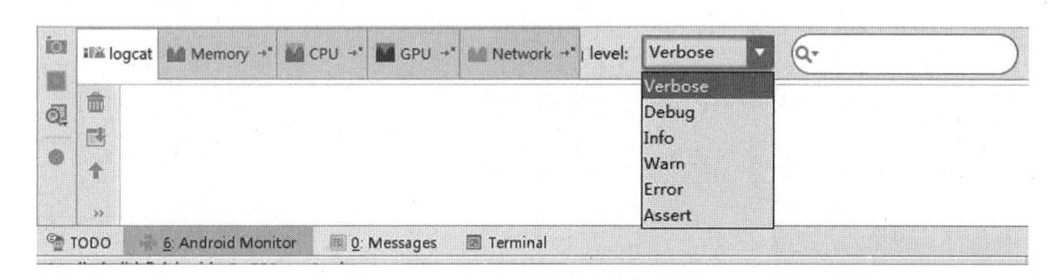

图 4－4　LogCat 窗口

LogCat 将输出的日志信息分成不同的级别，方便日志的分级筛选和查看。六个级别分别是：

- Verbose，详细模式。这个级别最低，该模式下，显示所有级别的信息。代码中采用 Log. v( )输出。
- Debug，调试模式。输出除 Verbose 之外的所有信息。代码中采用 Log. d( )输出。
- Info，信息模式。输出 Info、warn、error 信息。代码中使用 Log. i( )输出。
- Warn，警告模式。输出 warn、error 信息。代码中使用 Log. w( )输出。
- Error，错误模式。输出 error 信息。代码中使用 Log. e( )输出。
- Assert，断言模式。输出 Assert 信息。

由于日志信息量非常大，除了分级之外，LogCat 提供了过滤器来帮助我们筛选或过滤指定信息，可以自定义添加过滤器，在过滤器对日志信息标签(tag)、产生日志的进程编号(pid)、信息内容、包名以及上面的信息级别进行设置，即可对日志内容进行过滤，如图 4－5 所示。

在代码中调用 Log 类的响应方法即可输出相应级别的日志信息。例如，在 HelloWorld 项目中，Activity 创建时输出信息的代码如下：

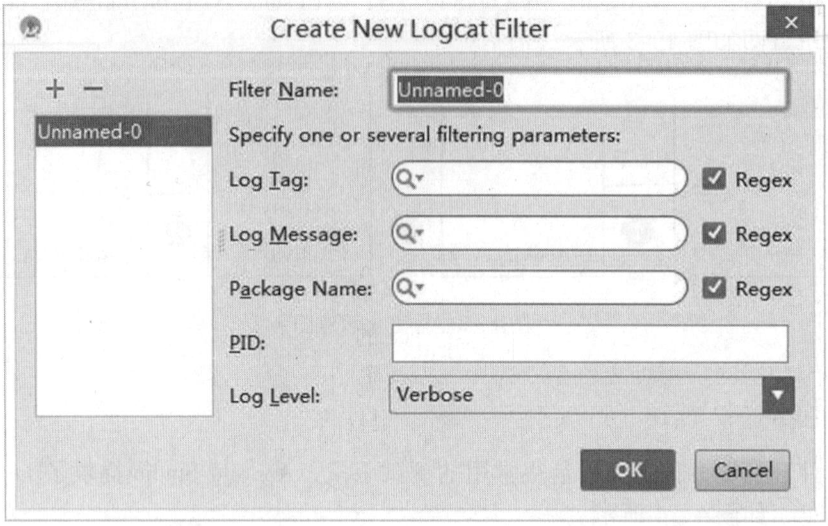

图 4－5　创建 LogCat 过滤器

```
public class MainActivity extends AppCompatActivity
{
      private static final String TAG = "LogTest";
      @ Override
      protected void onCreate(Bundle savedInstanceState) {
            super.onCreate(savedInstanceState);
            setContentView(R.layout.activity_main);
            Log.i(TAG,"onCreate");
      }
}
```

当代码运行时，onCreate 执行时会在 LogCat 窗口看到输出的信息文本，如图 4－6 所示。

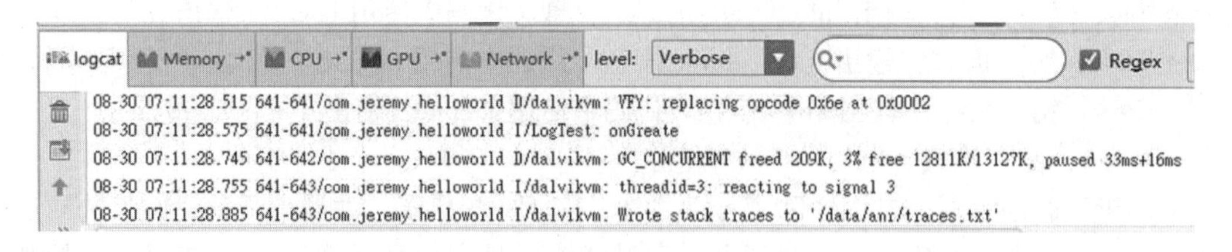

图 4－6　LogCat 日志信息

4.1.4　使用 LogCat 来验证 Activity 生命周期

在 Activity 生命周期的七个方法中，分别写上输出不同日志信息的代码，这样就可以在应用运行时观察七个方法调用的时间以及彼此之间的顺序。

4.2 Service、BroadcastReceiver、ContentProvider

除了 Activity 组件以外的三个组件，相对而言在游戏中应用较少，下面简单介绍一下。

1. Service

Activity 总是装载一个视图来与用户交互，而 Service(服务)是没有用户界面的程序，用来开发监控类程序。例如音乐播放器，一般的音乐播放器应用当然要包括 Activity 来与用户交互，用户选择播放或者下载删除等操作，但其中的音乐播放功能在应用转到后台时音乐应该保持播放，这就要通过 Service 组件来实现；还有通常的下载功能，也需要能在后台继续运行。

2. BroadcastReceiver

BroadcastReceiver(广播接收器)也没有 UI，不需要和用户交互，可以在后台运行。相对 Service 而言，BroadcastReceiver 负责短时间处理，该组件可以过滤并针对性地接收外部事件(例如来电)，并且对该事件进行响应。响应的方式可以是启动其他的组件(如 Activity 或 Service)来处理，或者也可以发送通知给用户。

3. ContentProvider

ContentProvider(内容提供者)用于在 Android 系统实现数据共享，使一个应用的数据可以提供给其他应用访问。例如通讯录、短信等等，都是 Android 系统提供的 ContentProvider，只要有权限，都可以在自己开发的应用中进行访问。

4.3 Context

Context(上下文)是一个抽象类，在 Android 应用开发中会在很多地方用到。它代表 App 的运行环境，所以要访问当前包的资源，或者要与组件进行通信，都需要用到它。在这个抽象类中定义了一套完成上述功能的接口，Activity 和 Service 都是实现这些接口的子类。除了这两个组件，还有一个 Application，即应用程序类。这三个类的实例也都可以通过 Context 的方法来访问资源，交互组件。

4.4 Intent

在 HelloWorld 项目中，只有一个 Activity 组件。实际上，一个完整的应用肯定会包括多个乃至多种组件，而各个组件之间的交互和通信则要通过 Intent(意图)来完成。组件通过 Intent 来向 Android 系统表达其某种请求或意愿。无论是跳转到其他组件，或是发送信息，实际上都是通过 Intent 请求系统完成的。四大组件中，ContentProvider 本身是数据共享，不需要通过 Intent 通信，另外三个组件接收 Intent 的方法如表 4 - 1 所示。

表 4 - 1　组件使用 Intent 通信方法表

Activity	startActivity	跳转 Activity
	startActivityForResult	跳转并关闭目标后返回
	setResult	设置返回的结果
Service	startService	启动服务
	bindService	绑定服务
BroadcastReceiver	sendBroadcast	发送广播
	sendOrderedBroadcast	发送有序广播
	sendStickyBroadcast	发送黏性广播

4.4.1　Intent 的两种方式

Intent 作为 Android 用于组件彼此调用和信息传递的机制，需要存放调用的行为和附加的数据作为 Intent 对象，包含 Components（目标组件）、Action（动作）、Data（数据）、Category（类别）、Extras（附加数据）、Flags（标志）这六个属性。

Intent 用于组件间的通信，目标组件可以直接指定目标组件名字来直接调用，也可以通过 Intent Filter（意图过滤器）来寻找目标组件。前一种方式称为显式调用，后一种称为隐式调用。例如，可以直接通过人的名字找到某人沟通，也可以通过年龄、籍贯、职业等人的属性找到合适的人进行沟通。提供的属性可能有多个组件满足，系统会给出一个列表供用户选择。例如，在某个应用中要访问某个站点，如果在系统中安装多个浏览器的话，会有提示询问用哪个浏览器打开。显示调用简单明了，但需要知道目标组件名，通常两个组件在同一个应用中的情况会更适合采用显式调用。如果要调用应用外的组件，也不知道目标组件名，就可以通过 Intent Filter，设置合适的属性进行隐式调用。

4.4.2　Intent

六个属性中 Components 用来指定 Intent（属性）的目标组件的名。如果 Intent 属性有指定，那么直接根据该值进行显式调用目标组件即可，其他所有属性都是可选的。设置该属性方法有四种：

- setComponent（ComponentName name），参数 ComponentName 是组件封装类，包括包名、类名或 Context 对象。
- setClassName（String packageName，String classNameInThatPackage），参数分别为包名和组件类名。
- setClassName（Context context，String classNameInThatContext），参数分别为上下文和目标组件类名。
- setClass（Context context，Class classObjectInThatContext），参数分别为上下文和目标组件类型。

例如，要从当前的 Activity 跳转到另一个 Activity，代码如下：

```
Intent intent = new Intent();
intent.setClass(this, OtherActivity.class);
startActivity(intent);
```

上述代码中，OtherActivity 是要跳转的目标 Activity 类名，第二行代码中的第一个参数 this 代表该代码所处当前 Activity 的对象实例。作为 Context 实例传入，这里创建 Intent 对象时调用了无参构造函数，实际上 Intent 也提供有参构造函数可以合并上述前两行代码：

```
Intent intent = new Intent(this, OtherActivity.class);
startActivity(intent);
```

而如果调用本应用之外的组件，无法获知类名，可以使用隐式调用。隐式调用要通过 Intent Filter 实现。隐式调用没有目标组件名，只能通过已知的其他属性去寻找合适的组件。Intent Filter 会根据 Intent 的 Action、Data 和 Category 在 Android 系统进行匹配，找到适合的组件来响应 Intent。想要得到匹配，目标组件在注册时必须设置 Intent Filter。例如，HelloWorld 项目中的 AndroidManifest.xml 中的代码就指明了其唯一组件 MainActivity 能够响应的 Intent Filter，代码如下：

```
<?xml version="1.0" encoding="utf-8"?>
<manifest xmlns:android="http://schemas.android.com/apk/res/android"
   package="com.jeremy.helloworld" >
    <application
        android:allowBackup="true"
        android:icon="@mipmap/ic_launcher"
        android:label="@string/app_name"
        android:supportsRtl="true"
        android:theme="@style/AppTheme" >
        <activity android:name=".MainActivity" >
              <intent-filter>
                    <action android:name="android.intent.action.MAIN"/>
                    <category android:name="android.intent.category.
                     LAUNCHER"/>
              </intent-filter>
        </activity>
    </application>
</manifest>
```

如上述代码所示，在自身应用的清单文件中完成注册并设置好 Intent Filter 才能被其他组件隐式调用，当然，要属性匹配。匹配的三个属性详细说明如下：

(1) Action(动作)。Action 指 Intent 要做的动作，它通过一个字符串常量来描述。Android 提供的通用的 Action，如表 4-2 所示。通过 setAction 和 getAction 设置和读取

Intent 的 Action 属性。

表 4 -2　Action 列表

常　量	目标组件	动　作
ACTION_ CALL	activity	启动电话
ACTION_ EDIT	activity	显示用户编辑的数据
ACTION_ MAIN	activity	开启一个任务，启动其初始 Activity
ACTION_ VIEW	activity	显示数据
ACTION_ SYNC	activity	同步服务器数据
ACTION_ BATTERY_ LOW	broadcast receiver	电量不足警告
ACTION_ HEADSET_ PLUG	broadcast receiver	耳机插入或拔出
ACTION_ SCREEN_ ON	broadcast receiver	打开屏幕
ACTION_ TIMEZONE_ CHANGED	broadcast receiver	时区设置变更

(2) Data(数据)。Data 指定了目标组件需要的数据，由指定数据的 URI 和 MIME 类型组成，不同的动作会对应携带不同的数据。

(3) Category(类别)。提供能够处理这个 Intent 对象的组件的类别。常用的类别如表 4 -3 所示。

表 4 -3　Category 列表

常　量	含　义
CATEGORY_ BROWSABLE	目标 Activity 可以安全调用浏览器显示数据
CATEGORY_ GADGET	目标 Activity 可以嵌入其他 Activity 里面
CATEGORY_ HOME	目标 Activity 是 Home 主画面
CATEGORY_ LAUNCHER	目标 Activity 是应用中的起始 Activity
CATEGORY_ PREFERENCE	目标 Activity 是偏好设置的 Activity

根据三个属性，可以设置 Intent Filter 来进行隐式调用，如果有匹配的组件将成为目标组件。例如，应用中需要访问网站，代码如下：

```
Intent i = new Intent(Intent.ACTION_VIEW);
    String uriStr = "http://www.sise.com";
    i.setData(Uri.parse(uriStr));
    this.startActivity(i);
```

还有两个属性。一个是 extras(附加信息)，顾名思义，这是对组件信息的补充。例如，要发送电子邮件，可以把邮件内容保存在 extras 属性中。另一个是 Flags(标志)，用于指定目标组件的任务行为，例如，开启一个新的任务或者清空 Activity 栈。

4.4.3 Activity 之间数据传递

如前所述，Intent 的 data 属性可以存放数据，extras 可以存放附加数据，两个属性都可以用来在 Activity 之间传递数据。我们在 HelloWorld 中添加一个新的 Activity 类，名为 SecondActivity，操作方是打开 file—New—Activity—Empty Activity，如图 4－7 所示。

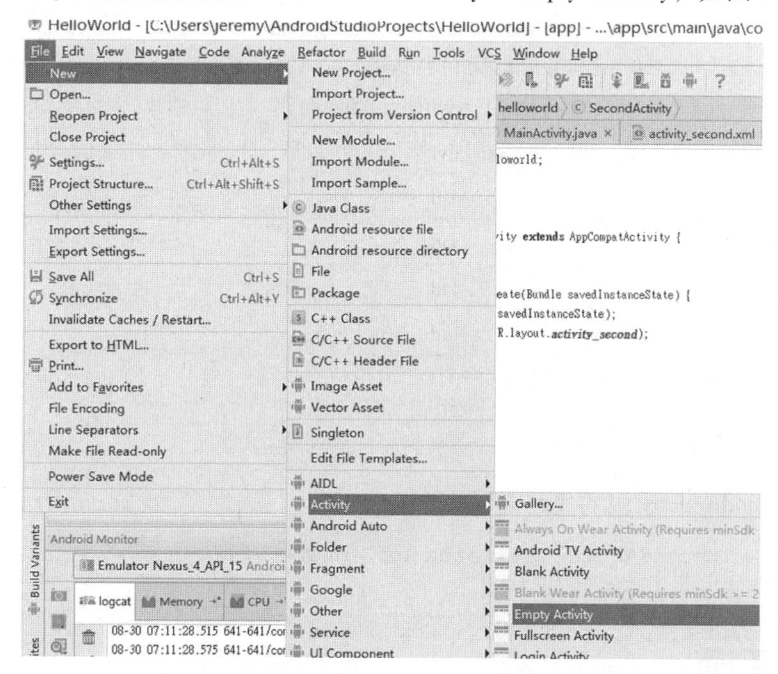

图 4－7　添加一个新的 Activity

在弹出的 Activity 设置界面写上 Activity 类名，点击“Finish”即可，如图 4－8 所示。

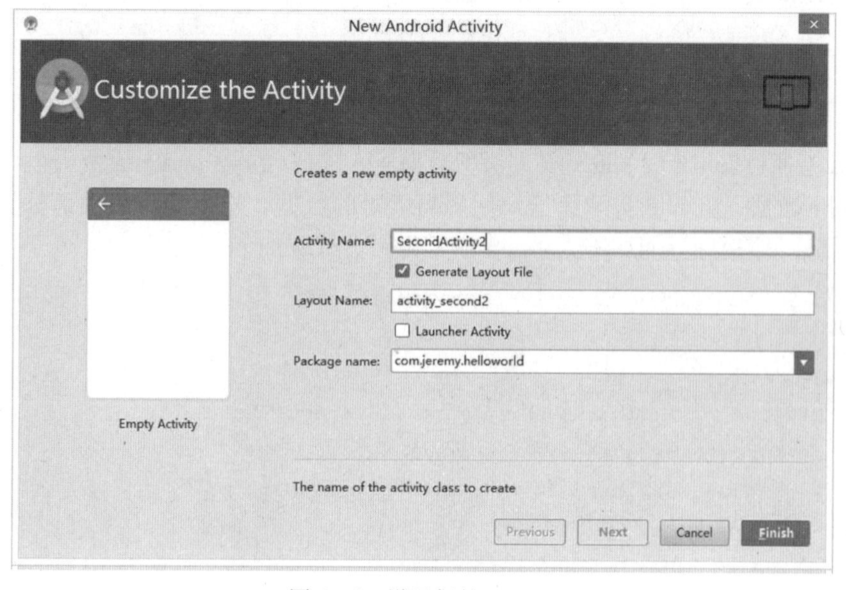

图 4－8　设置新的 Activity

用这种方法添加的 Activity，可以打开 AndroidManifest. xml，发现该 Activity 已经注册好了(切记，组件要在清单文件中注册才能访问)，并且有自己的布局，相当方便。在 MainActivity 中的代码如下：

```
public class MainActivity extends AppCompatActivity
{
    @ Override
    protected void onCreate(Bundle savedInstanceState) {
        super. onCreate(savedInstanceState);
        setContentView(R. layout. activity_main);

        Intent intent = new Intent();
        intent. setClass(this, SecondActivity. class);
        String uriStr = "http://www. sise. com";
        intent. setData(Uri. parse(uriStr));
        startActivity(intent);
    }
}
```

而在 SecondActivity 中的代码如下：

```
public class SecondActivity extends AppCompatActivity
{
    @ Override
    protected void onCreate(Bundle savedInstanceState) {
        super. onCreate(savedInstanceState);
        setContentView(R. layout. activity_second);
        Uri uri = getIntent(). getData();
        Log. i("second", uri. toString());
    }
}
```

extras 是通过 Intent 的 putExtra 方法把数据放入该属性的，putExtra 放入的参数可以采用 Bundle 数据，Bundle 实际是“Key - Value”对的数据。与上面代码大部分类似，使用 extras 属性只需要把 setData 改为 putExtras 即可，代码如下：

```
public class MainActivity extends AppCompatActivity
{
    @ Override
    protected void onCreate(Bundle savedInstanceState) {
        super. onCreate(savedInstanceState);
        setContentView(R. layout. activity_main);

        Intent intent = new Intent();
```

```
        intent. setClass (this, SecondActivity. class);
        Bundle bundle = new Bundle ();
        Bundle. putString ("url", "www. sise. com");
        intent. putExtra (bundle);
        startActivity (intent);
    }
}
```

而 SecondActivity 中读取该数据的代码改为：

```
public class SecondActivity extends AppCompatActivity
{
    @ Override
    protected void onCreate (Bundle savedInstanceState) {
        super. onCreate (savedInstanceState);
        setContentView (R. layout. activity_second);
        Bundle bundle = getIntent (). getExtra ();
        Log. i ("second", bundle. getString ("url"));
    }
}
```

5 UI 布局和控件

Activity 是四大组件唯一与用户交互的组件。Activity 通过加载一个 View(视图)来显示用户看到的图像。

5.1 View 和 ViewGroup

View 是一个超类，Android 中的布局和各类控件都是继承自 View 类。用户看到的应用界面是由一个或多个 View 组成的。Android 中 View 及其部分常用子类的类图如图 5－1所示。

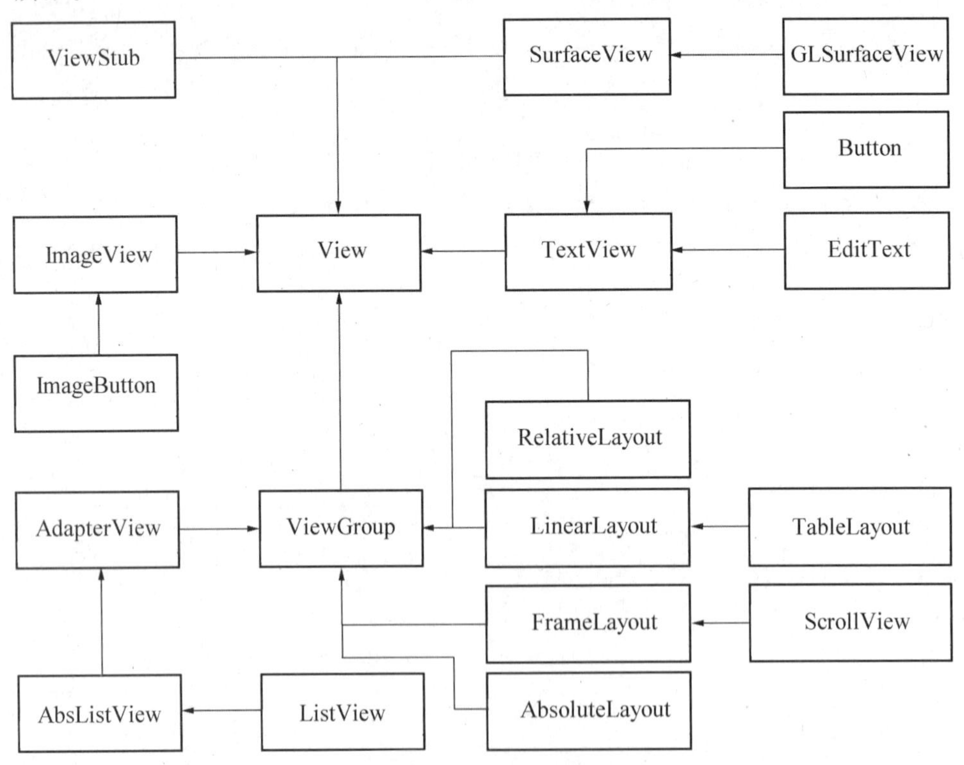

图 5－1 View 及其部分子类

从图 5－1 可以看出，View 的部分直接子类包括一部分简单控件和 ViewGroup 以及后面会述及的 SurfaceView，而 View 的子类 ViewGroup 是一个视图容器类，该类及其子类除了具有 View 的特性之外，其中可以包含其他的 View 和 ViewGroup，这样应用界面就可以由很多个 View 对象通过嵌套来实现复杂的图形界面，形成一个树形的层次结构，而所有的容器控件都是 ViewGroup 的子类，包括布局类。有了 ViewGroup，整个视图就可以由多个 View 组成丰富的层次结构，如图 5 – 2 所示。

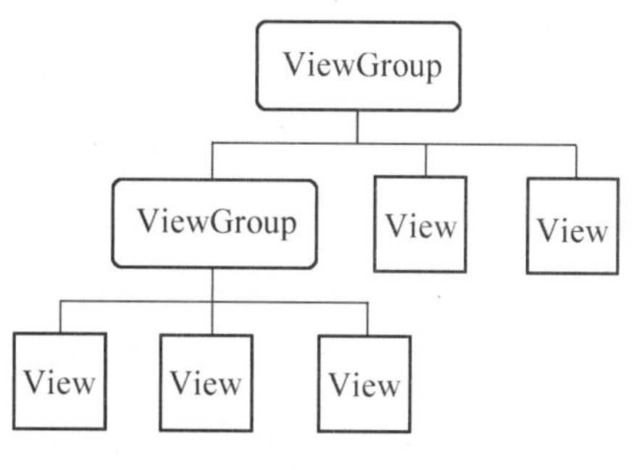

图 5 – 2　Android View 组织结构

5.2　控件

大多数的控件与其他语言和平台的控件属性方法都类似，限于篇幅，本书仅列举几款常用控件。

5.2.1　TextView

TextView 就是文本框控件，在其他语言中或称为 label。和 edit 不同，它不接受用户输入，只读性质，只负责显示文本信息。在 HelloWorld 项目中，就使用了一个 TextView 控件。

```
<TextView
        android:text = "Hello World!"
        android:layout_width = "wrap_content"
        android:layout_height = "wrap_content" />
```

该控件添加在 activity_main.xml 中，当该 .xml 文件被装载到 Activity 中时，TextView 就显示出来，并且它的 text 属性赋值为“HelloWorld”。如果需要在程序中访问该控件，那么该控件就需要一个 ID 值。将上述代码改为：

```
<TextView
    android:text = "Hello World!"
    android:id = "@ + id/TxtViewHello"
    android:layout_width = "wrap_content"
    android:layout_height = "wrap_content" />
```

在 XML 中，@ 符号代表引用资源，@ 后面接资源名称。例如@ string/hello，表示 string 下面有个名为 hello 的资源，而不是代表“hello”这个文本。@ 后面有个 + 号，表示原本没有这个资源标识，现要新添资源标识。因此，上面代码表示，在 R 文件(即资源标识文件)中添加了一个 ID 定义 TxtViewHello，该 ID 就标记着显示的 TextView 控件。有了 ID，就可以通过 AppCompatActivity 提供的 findViewById 这个方法来获得指定控件。获得控件的对象引用后可以通过 setText 方法修改其 text 的属性，即显示的文本内容。

```
TextView txtHello = (TextView)findViewById(R.id.txtViewHello);
txtHello.setText("Hello New World");
```

5.2.2 Button

Button 是 TextView 的子类，它具有 TextView 的属性方法，可以通过 setText 来修改其文本内容。除此之外，Button 主要是用来与用户进行交互，响应用户的点击事件。Android 的事件处理机制与 Java GUI 的事件处理是类似的，主要是由 Event Source(事件源)、Event(事件)以及 Event Listener(事件监听器)组成。例如，要在 HelloWorld 界面添加一个按钮，点击时更改 HelloWorld 的文本信息为“Hello New World”，首先就要在相应的布局文件中添加一个 Button 控件。

```
<RelativeLayout
    xmlns:android="http://schemas.android.com/apk/res/android"
    xmlns:tools="http://schemas.android.com/tools"
    android:layout_width="match_parent"
    android:layout_height="match_parent"
    android:paddingLeft="@dimen/activity_horizontal_margin"
    android:paddingRight="@dimen/activity_horizontal_margin"
    android:paddingTop="@dimen/activity_vertical_margin"
    android:paddingBottom="@dimen/activity_vertical_margin"
    tools:context="com.jeremy.helloworld.MainActivity">

    <TextView
        android:text="Hello World!"
        android:id="@+id/TxtViewHello"
        android:layout_width="wrap_content"
        android:layout_height="wrap_content"/>
    <Button
        android:text="NewWorld!"
        android:id="@+id/btnHello"
        android:layout_width="wrap_content"
        android:layout_height="wrap_content"
        android:layout_centerHorizontal="true"
        android:layout_centerVertical="true"/>
</RelativeLayout>
```

在上述代码中，相对于 TextView 控件，新加的 Button 控件多设置了两个属性，用于将其置于屏幕中央，和 TextView 的位置错开，避免遮挡。可以在 Activity 装载的布局中添加一个 Button，其上面显示的文本为“NewWorld!”，该控件的 ID 标记为 btnHello，为其添加点击事件处理代码如下：

```
protected void onCreate(Bundle savedInstanceState) {
        super.onCreate(savedInstanceState);
        setContentView(R.layout.activity_main);
        final TextView txtHello = (TextView)findViewById(R.id.TxtViewHello);
        Button btnHello = (Button) findViewById(R.id.btnHello);
        btnHello.setOnClickListener(new View.OnClickListener() {
            @Override
            public void onClick(View v) {
                  txtHello.setText("Hello New World!");
            }
        });
}
```

上述代码首先通过 findViewById 得到两个控件的引用。由于 findViewById 得到的是 View 类，因此要转换成 TextView 和 Button，然后给事件源 txtHello 添加一个 Click 事件的监听，并新建一个 onClickListener 来监听该事件，重写其中的 onClick 方法进行对事件监听后的处理。

5.2.3 ImageView

ImageView 是显示一张图片的控件，与 TextView 相比，一个显示文字，另一个显示图片。设置图片的方法原型为 void setImageBitmap(Bitmap bitmap)。

5.2.4 ListView

ListView(列表框)继承自 ViewGroup，所以它可以包含其他的 View 子类对象。ListView 的使用比前述的简单控件复杂，因为它显示的数据量和功能更多，使用它需要指定 Adapter(适配器)和数据源。数据源的形式很多，可以是数据库、XML 文件或数组、集合。而 Adapter 则是 ListView 与这些数据源的中介，负责将不同的数据绑定到控件上。

5.2.5 Toast

Toast 是 Android 提供的提示框控件，负责给用户显示一些帮助或提示信息，这些信息可以是文字也可以是图像。与接触比较多的 Dialog(对话框)相比，它没有焦点，并在显示一段时间之后会自动消失。下面是将之前点击按钮后改变 txtView 的文本改为按钮点击后显示一个 Toast 的代码。

```
protected void onCreate(Bundle savedInstanceState) {
        super.onCreate(savedInstanceState);
        setContentView(R.layout.activity_main);
        final TextView txtHello = (TextView) findViewById(R.id.TxtViewHello);
```

```
    Button btnHello = (Button) findViewById(R.id.btnHello);
    btnHello.setOnClickListener(new View.OnClickListener() {
        @Override
        public void onClick(View v) {
            Toast.makeText(MainActivity.this,"Hello New World!",
                                         Toast.LENGTH_LONG).show();
        }
    });
}
```

Toast.makeText 会创建一个 Toast 对象，并在其中参数设置显示文本和显示持续时间，然后通过 show 方法显示出来。

5.3 布局

ViewGroup 的子类中，除了容器控件外，就是布局类。布局表明用户界面的架构，负责设置各个 UI 元素在屏幕上的位置，Android 中有以下五大布局：

- FrameLayout(框架布局)。最简单的布局格式，也称单帧布局。
- LinearLayout(线性布局)。所有 UI 元素在单一方向依次排列，比如垂直或水平排列。
- AbsoluteLayout(绝对布局)。所有 UI 元素通过自身坐标指定在屏幕中的位置。
- RelativeLayout(相对布局)。UI 元素可以指定其相对于其他元素的位置。
- TableLayout(表格布局)。通过表格将 UI 元素指定行列放置。

布局之间还可以嵌套使用，形成更为复杂的 UI 界面，Layout 可以通过 XML 来设置。例如，HelloWorld 项目中的 activity_main.xml 文件，其中用到了 RelativeLayout 相对布局。在 XML 中，可以看到很多名为 layout_something 的 XML 布局属性，这是为 UI 元素定义相应的 ViewGroup 的布局参数。由于 ViewGroup 可以包含其他的 View 类对象，因此每个 View 通过设置 layout_something 来指定自身在父容器中的位置和大小。

5.3.1 FrameLayout

FrameLayout 是最简单的一种布局方式，所有的子元素固定在屏幕的左上角显示，无法直接指定特定位置。如果有多个 UI 元素，则会叠加在一起。要改变位置，可通过 layout_gravity 设置靠哪个方位放，如 left | center_vertical 是左侧中间，right | bottom 是靠右下，整个屏幕区域分成九宫格，左上、左中、左下、右上、右中、右下、上中、中、下中。在这个基础上还可以通过 layout_margin 设置靠边的距离，例如 layout_gravity 设置 right | bottom，即靠右下放置，这时将 layout_marginright 设为 100dp，即靠右侧边 100dp 的位置。单帧布局极其简单，通常会和其他布局搭配使用。单帧布局效果如图5－3所示，其中 XML 第一个 imagView 是最小的，第二个是居中

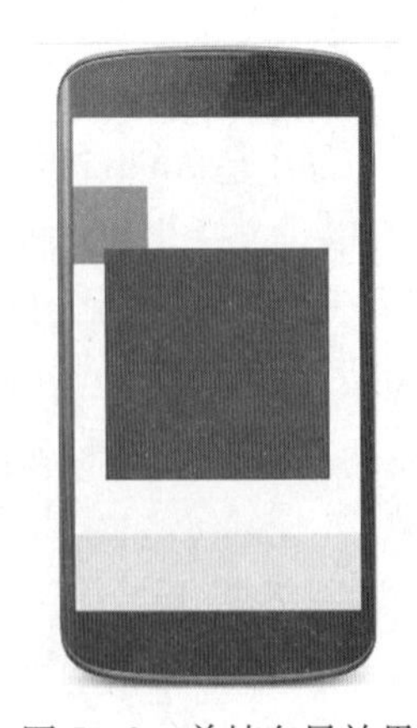

图 5－3　单帧布局效果

的，第三个是底部颜色最浅的。

相应的 . xml 文件代码如下：

```
<FrameLayout xmlns:android="http://schemas.android.com/apk/res/android"
    android:layout_width="match_parent"
    android:layout_height="match_parent">

    <ImageView
        android:layout_width="wrap_content"
        android:layout_height="wrap_content"
        android:id="@+id/imageView"
        android:layout_gravity="left|center_vertical"
        android:background="#646464"
        android:maxHeight="100dp"
        android:maxWidth="100dp"
        android:minHeight="100dp"
        android:minWidth="100dp"
        android:layout_marginRight="100dp"
        android:layout_marginTop="20dp"
        android:layout_marginBottom="200dp" />

    <ImageView
        android:layout_width="wrap_content"
        android:layout_height="wrap_content"
        android:id="@+id/imageView2"
        android:layout_gravity="center"
        android:background="#323232"
        android:maxHeight="300dp"
        android:maxWidth="300dp"
        android:minHeight="300dp"
        android:minWidth="300dp"
        android:nestedScrollingEnabled="true" />

    <ImageView
        android:layout_width="wrap_content"
        android:layout_height="wrap_content"
        android:id="@+id/imageView3"
        android:layout_gravity="center_horizontal|bottom"
        android:maxHeight="100dp"
        android:maxWidth="600dp"
        android:minHeight="100dp"
        android:minWidth="600dp"
        android:background="#c8c8c8" />

</FrameLayout>
```

5.3.2 LinearLayout

线性布局 LinearLayout 比较常用，子元素在其中依次按垂直或水平的方向排列，每一个子元素都位于前一个元素之后。如果是垂直排列，那么将是一个 *N* 行单列的结构，每一行有一个元素。如果是水平排列，那么将是一个单行 *N* 列的结构。搭建两行两列的结构，通常的方式是先垂直排列两个元素，每一个元素里再包含一个 LinearLayout 进行水平排列，效果如图 5－4 所示。该布局只有一个线性布局，所有三个元素只能在列上次序排列，第二个左侧和第一个右侧对齐，第三个左侧和第二个右侧对齐。

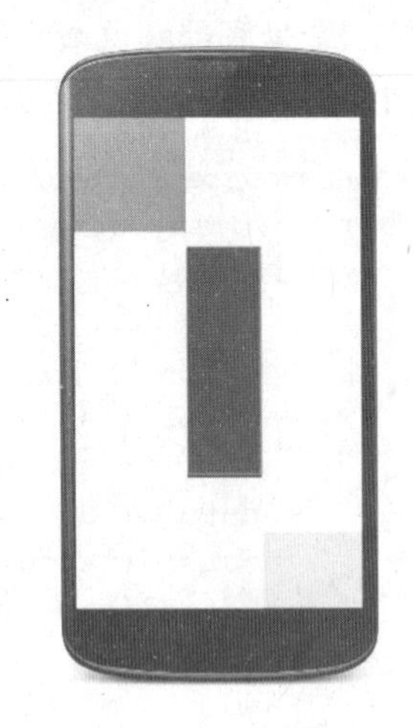

图 5－4　线性布局效果

相应的 .xml 文件代码如下：

```
<LinearLayout xmlns:android="http://schemas.android.com/apk/res/android"
    android:layout_width="match_parent"
    android:layout_height="match_parent">

    <ImageView
        android:layout_width="wrap_content"
        android:layout_height="wrap_content"
        android:id="@+id/imageView4"
        android:background="#787878"
        android:maxHeight="150dp"
        android:maxWidth="150dp"
        android:minHeight="150dp"
        android:minWidth="150dp"
        android:nestedScrollingEnabled="false" />

    <ImageView
        android:layout_width="wrap_content"
        android:layout_height="wrap_content"
        android:id="@+id/imageView5"
        android:layout_gravity="center_vertical"
        android:maxHeight="300dp"
        android:maxWidth="100dp"
        android:minHeight="300dp"
        android:minWidth="100dp"
        android:background="#323232" />
```

```
    <ImageView
        android:layout_width="wrap_content"
        android:layout_height="wrap_content"
        android:id="@+id/imageView6"
        android:layout_gravity="bottom"
        android:maxHeight="100dp"
        android:maxWidth="600dp"
        android:minHeight="100dp"
        android:minWidth="600dp"
        android:background="#c8c8c8" />
</LinearLayout>
```

5.3.3 AbsoluteLayout

绝对布局 AbsoluteLayout 直接指定 UI 元素的 x、y 坐标值，并显示在屏幕上，屏幕坐标系的原点在屏幕左上角，横坐标正方向向右，纵坐标正方向向下。绝对布局很容易理解，但实际应用不多，因为使用绝对坐标，在不同的设备上或需要支持屏幕旋转时，界面很容易出问题，效果如图 5－5 所示。

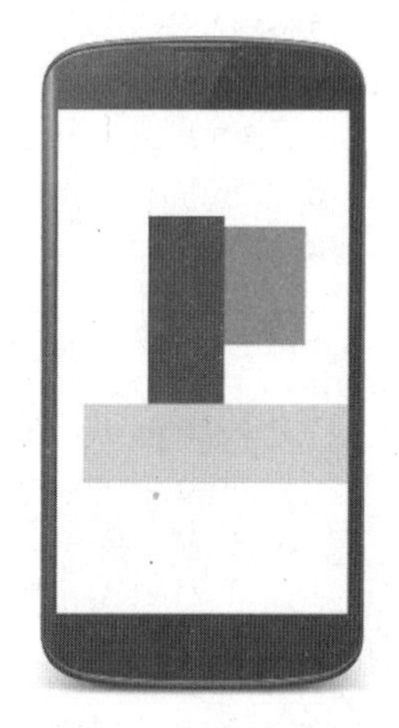

图 5－5　绝对布局效果

相应的 .xml 文件代码如下：

```
<AbsoluteLayout xmlns:android="http://schemas.android.com/apk/res/android"
    android:layout_width="match_parent"
    android:layout_height="match_parent">

    <ImageView
        android:layout_width="wrap_content"
        android:layout_height="wrap_content"
        android:id="@+id/imageView7"
        android:background="#787878"
        android:maxHeight="150dp"
        android:maxWidth="150dp"
        android:minHeight="150dp"
        android:minWidth="150dp"
        android:nestedScrollingEnabled="false"
        android:layout_x="175dp"
```

```
        android:layout_y="149dp" />

    <ImageView
        android:layout_width="wrap_content"
        android:layout_height="wrap_content"
        android:id="@+id/imageView8"
        android:layout_gravity="center_vertical"
        android:maxHeight="300dp"
        android:maxWidth="100dp"
        android:minHeight="300dp"
        android:minWidth="100dp"
        android:background="#323232"
        android:layout_x="118dp"
        android:layout_y="136dp" />

    <ImageView
        android:layout_width="wrap_content"
        android:layout_height="wrap_content"
        android:id="@+id/imageView9"
        android:layout_gravity="bottom"
        android:maxHeight="100dp"
        android:maxWidth="600dp"
        android:minHeight="100dp"
        android:minWidth="600dp"
        android:background="#c8c8c8"
        android:layout_x="34dp"
        android:layout_y="374dp" />
</AbsoluteLayout>
```

5.3.4 RelativeLayout

相对布局 RelativeLayout 允许子元素指定它们相对于其他元素或父元素的位置(通过 ID 指定)。可以左右上下对齐或置于屏幕中央的形式来排列两个元素。元素按顺序排列，因此，如果第一个元素在屏幕的中央，那么相对于这个元素的其他元素将以屏幕中央(第一个元素所处位置)的相对位置来排列。在图 5－6 中，最上面的图片框 1 相对屏幕定位，在其左侧的图片框 2 相对于 1 进行定位，最下面的图片框 3 则相对于 2 进行定位，相对布局中常用到的布局属性如表 5－1 所示。

相对布局效果如图 5－6 所示。

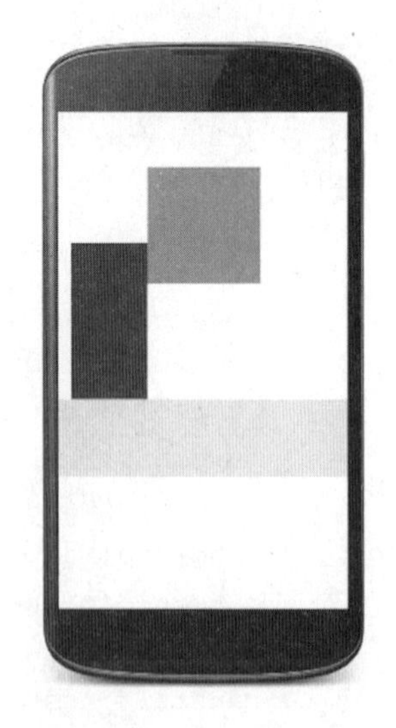

图 5－6　相对布局效果

表 5-1 布局中常用到的布局属性

布局属性	含 义
android：layout_ toLeftOf	该组件位于引用组件的左方
android：layout_ toRightOf	该组件位于引用组件的右方
android：layout_ above	该组件位于引用组件的上方
android：layout_ below	该组件位于引用组件的下方
android：layout_ alignParentLeft	该组件是否对齐父组件的左端
android：layout_ alignParentRight	该组件是否对齐父组件的右端
android：layout_ alignParentTop	该组件是否对齐父组件的顶部
android：layout_ alignParentBottom	该组件是否对齐父组件的底部
android：layout_ centerInParent	该组件是否相对于父组件居中
android：layout_ centerHorizontal	该组件是否横向居中
android：layout_ centerVertical	该组件是否垂直居中

相应的.xml文件代码如下：

```
< RelativeLayout xmlns: android = " http: // schemas. android. com/apk/res/
android"
    android:layout_width = "match_parent"
    android:layout_height = "match_parent" >

    < ImageView
        android:layout_width = "wrap_content"
        android:layout_height = "wrap_content"
        android:id = "@ + id/imageView10"
        android:background = "#787878"
        android:maxHeight = "150dp"
        android:maxWidth = "150dp"
        android:minHeight = "150dp"
        android:minWidth = "150dp"
        android:nestedScrollingEnabled = "false"
        android:layout_x = "175dp"
        android:layout_y = "149dp"
        android:layout_alignParentTop = "true"
        android:layout_centerHorizontal = "true"
```

```
        android:layout_marginTop="72dp" />

    <ImageView
        android:layout_width="wrap_content"
        android:layout_height="wrap_content"
        android:id="@+id/imageView11"
        android:layout_gravity="center_vertical"
        android:maxHeight="300dp"
        android:maxWidth="100dp"
        android:minHeight="300dp"
        android:minWidth="100dp"
        android:background="#323232"
        android:layout_x="118dp"
        android:layout_y="136dp"
        android:layout_centerVertical="true"
        android:layout_toLeftOf="@+id/imageView10"
        android:layout_toStartOf="@+id/imageView10" />

    <ImageView
        android:layout_width="wrap_content"
        android:layout_height="wrap_content"
        android:id="@+id/imageView12"
        android:layout_gravity="bottom"
        android:maxHeight="100dp"
        android:maxWidth="600dp"
        android:minHeight="100dp"
        android:minWidth="600dp"
        android:background="#c8c8c8"
        android:layout_x="34dp"
        android:layout_y="374dp"
        android:layout_alignBottom="@+id/imageView11"
        android:layout_alignParentRight="true"
        android:layout_alignParentEnd="true" />
</RelativeLayout>
```

5.3.5 TableLayout

表格布局 TableLayout 将子元素的位置分配到行或列中。一个 TableLayout 由许多的 TableRow 组成，但与想象不同的是，不能指定列，每行有多个单元，单元里面存放 UI 元素。表格由列和行组成许多的单元格。表格允许单元格为空。单元格不能跨列，这与 HTML 中的不一样。表格布局效果如图 5-7 所示，三个元素分别放在第 2 行第 2 列、第 4 行第 4 列、第 6 行第 6 列。实际上没有放置控件的行列会自动收缩，所以看起来第二个和第三个是相邻的，但也可以直接设置某行的 minHeight。例如，这里的第一行和第三行，第一个元素和第二个隔着的第三行因为指定了 minHeight，就没有收缩，所以这两个元素纵向上有距离。

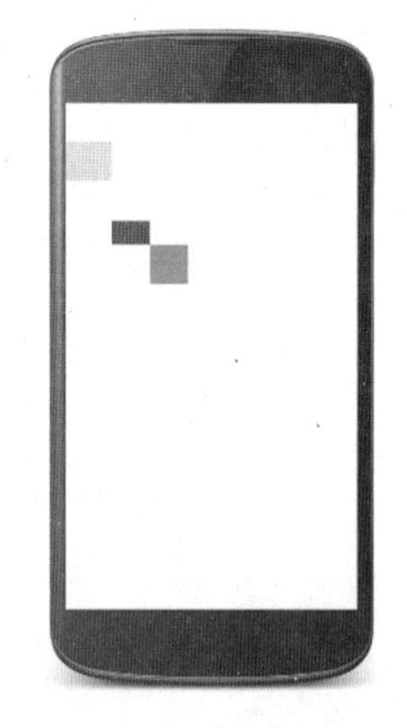

图 5-7　表格布局效果

相应的 .xml 文件代码如下：

```
<TableLayout xmlns:android="http://schemas.android.com/apk/res/android"
    android:layout_width="match_parent"
    android:layout_height="match_parent">

    <TableRow
        android:layout_width="match_parent"
        android:layout_height="wrap_content"
        android:minHeight="50dp"
        android:nestedScrollingEnabled="true">

    </TableRow>

    <TableRow
        android:layout_width="match_parent"
        android:layout_height="match_parent">

        <ImageView
            android:layout_width="wrap_content"
            android:layout_height="wrap_content"
            android:id="@+id/imageView15"
            android:layout_gravity="bottom"
            android:maxHeight="50dp"
            android:maxWidth="60dp"
```

```
            android:minHeight="50dp"
            android:minWidth="60dp"
            android:background="#c8c8c8"
            android:layout_x="34dp"
            android:layout_y="374dp"
            android:layout_column="2" />

    </TableRow>

    <TableRow
        android:layout_width="match_parent"
        android:layout_height="match_parent"
        android:minHeight="50dp"> </TableRow>

    <TableRow
        android:layout_width="match_parent"
        android:layout_height="match_parent">

        <ImageView
            android:layout_width="wrap_content"
            android:layout_height="wrap_content"
            android:id="@+id/imageView14"
            android:layout_gravity="center_vertical"
            android:maxHeight="30dp"
            android:maxWidth="50dp"
            android:minHeight="30dp"
            android:minWidth="50dp"
            android:background="#323232"
            android:layout_x="118dp"
            android:layout_y="136dp"
            android:layout_column="4" />
    </TableRow>

    <TableRow
        android:layout_width="match_parent"
        android:layout_height="match_parent">

    </TableRow>

    <TableRow
        android:layout_width="match_parent"
        android:layout_height="match_parent">

        <ImageView
            android:layout_width="wrap_content"
            android:layout_height="wrap_content"
            android:id="@+id/imageView13"
            android:background="#787878"
```

```
            android:maxHeight = "50dp"
            android:maxWidth = "50dp"
            android:minHeight = "50dp"
            android:minWidth = "50dp"
            android:nestedScrollingEnabled = "false"
            android:layout_x = "175dp"
            android:layout_y = "149dp"
            android:layout_column = "10" / >
    < /TableRow >

    < TableRow
        android:layout_width = "match_parent"
        android:layout_height = "match_parent" > < /TableRow >

    ……行数太多，此处省略部分代码……

    < TableRow
        android:layout_width = "match_parent"
        android:layout_height = "match_parent" >

    < /TableRow >
< /TableLayout >
```

在 Android Studio，要添加一个布局，非常简单，右键点击 res 的 layout，选择新增 XML 文件，就可以选择所选用的布局，如图 5 - 8 所示。

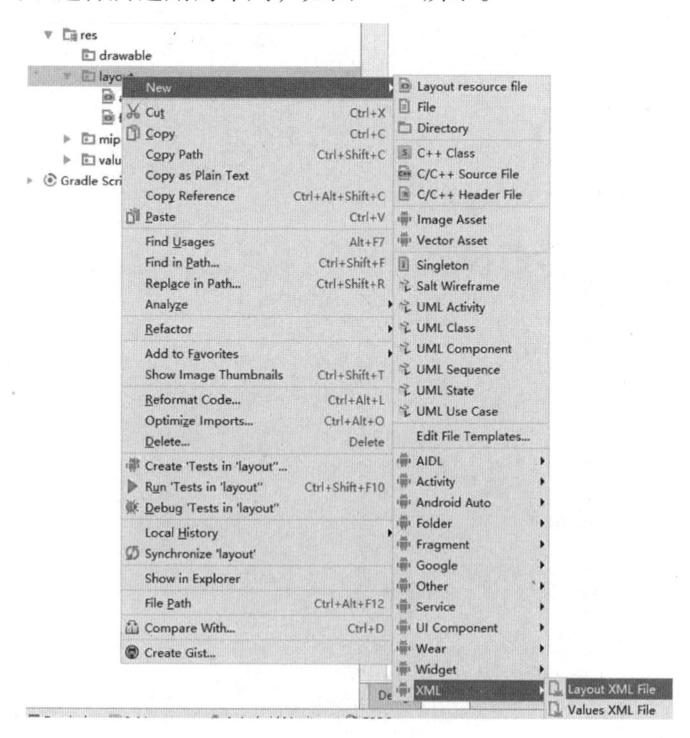

图 5 - 8　新建 Layout XML 文件

在弹出的创建 Layout 布局文件窗口可以输入 Layout 的文件名，以及想使用的基础布局。

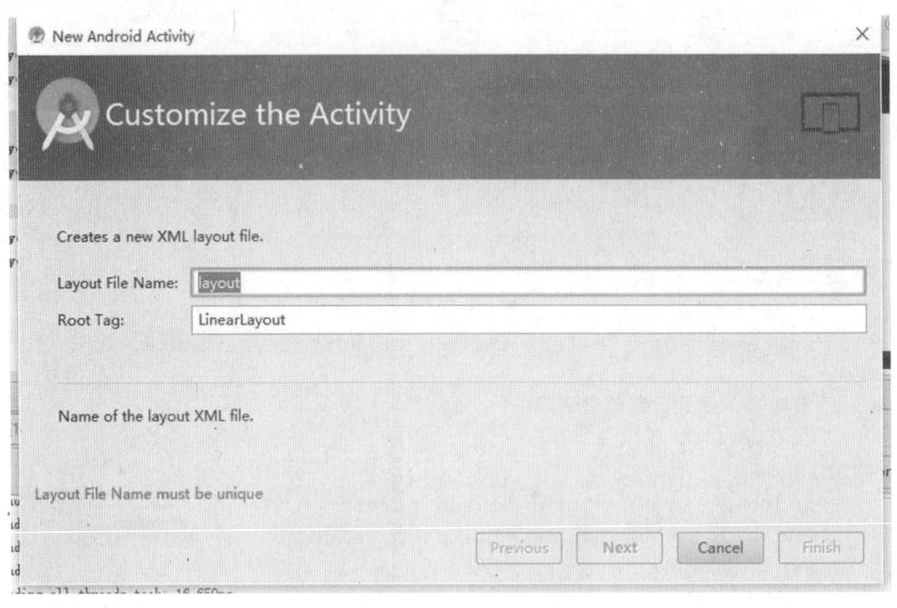

图 5－9　新建布局界面

创建布局文件后，会自动打开该文件，进入布局设计界面。布局文件可以用设计和文本两种方式查看，在可视化的设计界面可以直接拖放控件到布局中，拖放时根据位置不同会有布局属性的提示，拖放之后也可以在右侧的属性窗口直接修改各项布局属性，如图 5－10 所示。

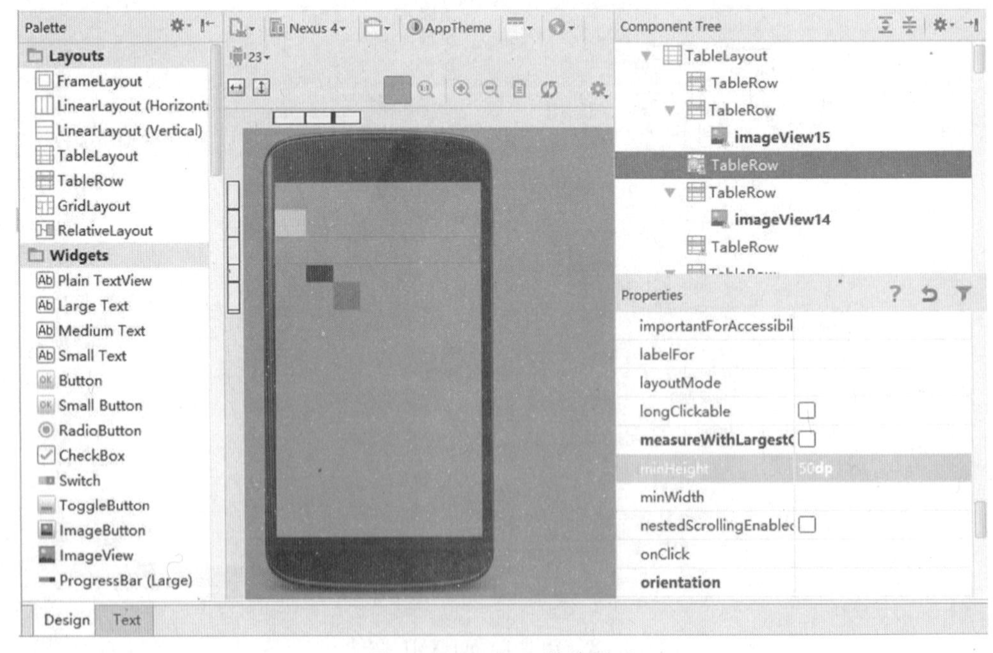

图 5－10　Layout 设计界面

在文本属性，可以在修改代码的同时，即时在右侧的 Preview 预览窗口看到改动的效果。

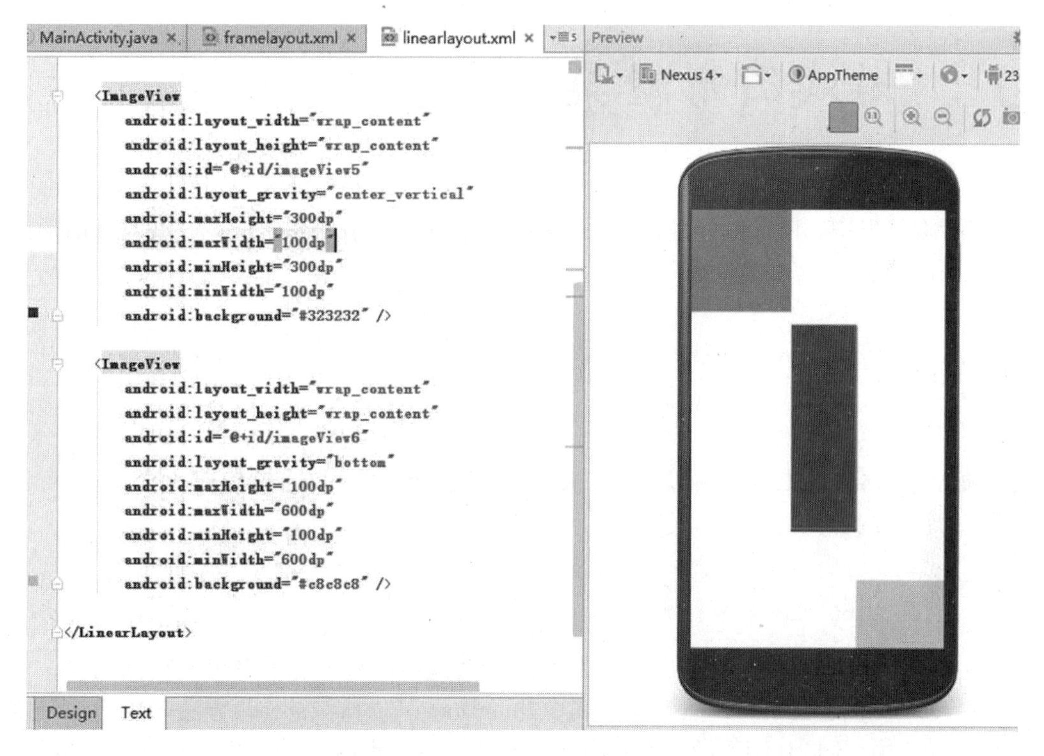

图 5－11 Layout 代码编辑界面

6 游戏图形渲染

在游戏开发中，图形编程是很重要的部分。相对于一般应用软件，游戏对于图形图像有更高的要求。

6.1 View

Activity 装载的视图除了用 XML 定义的布局之外，还可以直接构建一个 View 对象装载作为视图内容。例如下面代码：

```
View gameView = new View(this);
setContentView(gameView);
```

当然，由于 View 本身是空白的，因此没什么显示出来。为了自定义视图内容，可以继承 View 类创建其子类，然后重写其中的 onDraw(Canvas canvas)方法来决定绘制内容。Android 中绘制 2D 图像的 API 和类均在 android. graphics 包中，主要包括 Canvas(画布)、Bitmap(位图)、Paint(画笔)以及 Point、Rect 和 Color 等一些类组成。

6.1.1 Canvas

Canvas 是 View 所带的画布，需要在视图显示的内容必须画在画布上面。Canvas 位于 android. graphics 包中，是 Android 系统 2D 图形系统最核心的类，其中包括一系列的绘制方法，例如绘制各种几何图形、文本以及位图。下面代码是绘制文本：

```
Paint paint = new Paint();
protected void onDraw(Canvas canvas)
{
    paint.setColor(Color.WHITE);
    canvas.drawText("Game", textX, textY, paint);
}
```

从上述代码可以看出，重写 View 类的 onDraw 方法时，可以通过参数得到 Canvas 对象，从而进行绘图。上面代码调用了 drawText 进行文本的绘制，其他一些常用的方法如表 6 - 1 所示。

表 6－1 Canvas 绘图方法

方法原型	功 能
void drawARGB(int a, int r, int g, int b)	某颜色填充
void drawPoints(float[ ] pts, Paint paint)	绘制点
void drawLines(float[ ] pts, Paint paint)	绘制线
void drawRect(RectF rect, Paint paint)	绘制矩形
void drawCircle(float cx, float cy, float radius, Paint paint)	绘制圆
void drawArc(RectF oval, float startAngle, float sweepAngle, Boolean useCenter, Paint paint)	绘制弧形
void drawBitmap(Bitmap bitmap, Matrix matrix, Paint paint)	绘制位图

6.1.2 Bitmap

Bitmap(位图)是 Android 图像处理的重要类，游戏中最常用的资源。用来获取图像数据并渲染，进行图像的剪切、旋转、缩放等操作。Bitmap 实现在 android. graphics 包中。但是要注意 Bitmap 类的构造函数是私有的，外面并不能实例化，需要通过 BitmapFactory 类来创建。

利用 BitmapFactory 可以从一个指定文件中，利用 decodeFile( )解出 Bitmap；也可以从定义的图片资源中，利用 decodeResource( )解出 Bitmap。

```
Bitmap bmp =BitmapFactory.decodeResource(this.getResources(),
                                          R.drawable.img);
```

渲染一张位图，要定义一个新的类 GameView 继承自 View，在构造函数中从项目资源中加载位图，然后重写 onDraw 方法，在其中通过 Canvas 的 drawBitmap 将位图渲染出来。下面是渲染一张位图的代码。

```
public class GameView extends View {
     private Paint paint;
     private Bitmap bmp;
     private Matrix mat;
     public GameView(Context context)
     {
          super(context);
          paint = new Paint();
          mat = new Matrix();
          bmp =
BitmapFactory.decodeResource(context.getResources(),R.drawable.title);
 }
```

```
    @ Override
    protected void onDraw(Canvas canvas){
        super.onDraw(canvas);
        canvas.drawBitmap(bmp,mat,paint);
    }
}
```

上面代码绘制了一张位图，值得一提的是，drawBitmap 的第二个参数是 android.graphics.Matrix 类，代表用于位图变换的矩阵，通过不同的矩阵可以实现位图的 Rotate(旋转)、Translate(平移)、Scale(缩放)和 Skew(错切变换)。然后可以通过矩阵间的乘法得到复合矩阵，不同功能的矩阵相乘得到复合矩阵，具有所有参与相乘的矩阵的功能。例如，一个平移矩阵乘以旋转矩阵就得到先平移再旋转的复合矩阵。矩阵相关的方法名由相乘顺序(前缀)加上矩阵功能(后缀)组成，如表6－2所示。

表6－2　矩阵操作方法

	rotate	translate	skew	scale
pre	preRotate	preTranslate	preSkew	preScale
set	setRotate	setTranslate	setSkew	setScale
post	postRotate	postTranslate	postSkew	postScale

表6－2中，方法名前缀为 pre 表示先乘，即右乘，而 post 则相反。注意，矩阵的乘法不满足交换律，所以不同的相乘顺序得到的复合矩阵是不一样的。set 前缀则直接设置调用的矩阵对象的值，不进行相乘。举个例子，代码如下：

```
mat.postRotate(45f);
mat.postTranslate(100, 100);
mat.preScale(1.5f,1.5f);
mat.setTranslate(50,50);
```

当构建一个 Matrix 对象，调用其无参构造函数，会得到一个单位矩阵(不进行变换)。如果执行上述第一行代码，则在单位矩阵上后乘(左乘)，此时 mat 存放一个旋转45°矩阵；再执行第二行代码，同样后乘，则 mat 中存放先旋转再完成平移变换的复合矩阵，执行第三行代码后，注意方法是 pre 前缀，复合矩阵先乘缩放矩阵，那么就是先缩放再旋转最后平移的复合矩阵；然后再执行最后一条语句，由于是 set 前缀，因此平移(50，50)的矩阵将之前矩阵数据覆盖，最后 mat 中只是一个平移(50，50)的矩阵。

除了可以通过矩阵来实现位图的变换，Canvas 还提供了一系列方法来实现画布的变换。画布的变换会影响到所有绘制在该画布上的可视对象，所以使用画布变换要记得变换之后恢复画布。

6.1.3　View 绘图

用 View 类来开发游戏，因为游戏中每帧画面都需要更新，要注意 View 类中负责绘图的 onDraw 方法不能被直接调用。需要更新画面时要通过 view. invalidate( )方法请求 Android 重新绘制当前视图，也就是间接导致 onDraw 调用。代码如下：

```
@ Override
public boolean onTouchEvent (MotionEvent event) {
    switch (event. getAction ()) {
    case MotionEvent. ACTION_DOWN:
        return true;
    case MotionEvent. ACTION_UP:
        paint. setColor (Color. RED);
                this. invalidate ();
        return true;
    }
    return super. onTouchEvent (event);
}
```

注意，invalidate 只能工作在 UI 主线程，目前项目中只有一个主线程，当项目中创建了子线程，在其中调用 invalidate 时将会报错。此时可以通过另外一个方法 postInvalidate 来通知主线程更新 View 视图。

6.2 SurfaceView

View 更新 invalidate 必须在 UI 线程调用，但 UI 线程负担的工作比较多，棋牌类等更新频率比较低的游戏还可以采用 View，但要求较高的实时渲染的游戏则不适合。对于大多数渲染效率要求较高的游戏而言，更适合的是 SurfaceView。SurfaceView 是 View 的子类，内部实现了双缓冲机制，并且可以在子线程主动更新视图，使开发更为方便。

6.2.1 SurfaceHolder 成员

在 SurfaceView 类中有两个 Surface 成员，Surface 就是表面，其中自带了一个 Canvas 画布，所以 Surface 其实对应一块内存区域。不同于 View 只能被动地通知 UI 主线程更新，SurfaceView 可以随时锁定自身 Surface 中的画布来进行绘图，完成之后解锁并提交通知主线程更新。为了方便 SurfaceView 对 Surface 进行控制，android. view 包中提供了一个 SurfaceHolder 接口。每个 SurfaceView 对象都有一个 SurfaceHolder 成员，可以通过 getHolder 方法得到。相关代码如下：

```
public class GameSurfaceView extends SurfaceView implements
SurfaceHolder. Callback{
{
public GameSurfaceView (Context context)
{
```

```
        super(context);
        try {
            SurfaceHolder holder = getHolder();
            holder.addCallback(this);
        }catch (Exception e)
        {
        }
    }
    @ Override
    public void surfaceCreated(SurfaceHolder holder) {

    }

    @ Override
    public void surfaceChanged(SurfaceHolder holder, int format, int width, int
    height) {

    }

    @ Override
    public void surfaceDestroyed(SurfaceHolder holder) {

    }
}
```

上述代码定义了 GameSurfaceView 继承自 SurfaceView，并且还需要实现一个接口，这个接口是 SurfaceHolder. Callback。该接口用于监测 Surface 的工作状态，接口共三个方法：

- public void surfaceChanged(SurfaceHolder holder, int format, int width, int height){}
  该方法在 Surface 的大小发生改变时激发，例如横竖屏切换。
- public void surfaceCreated(SurfaceHolder holder){}　该方法在创建时激发，一般在这里加载图片资源、启动渲染的线程。
- public void surfaceDestroyed(SurfaceHolder holder){}　该方法在 Surface 销毁时激发，一般在这里将渲染的线程停止、图片资源释放等。

因为此处 GameSurfaceView 实现了 Callback 接口，所以给 SurfaceHolder 指定 this 为回调监听对象。游戏需要渲染的工作比较多，要新开一个线程来进行渲染的工作。

6.2.2 SurfaceView 子类

要使用 SurfaceView 渲染，首先定义一个类继承 SurfaceView，然后实现上节的代码（设置回调，实现三个方法），然后定义一个渲染方法。该方法与 View 的 onDraw 不同，它会在线程中主动调用，然后再定义一个渲染线程类来不断调用 SurfaceView 子类的渲染方法。代码如下：

```
public class GameSurfaceView extends SurfaceView implements SurfaceHolder.Callback{
    private RenderThread renderThread;
    private Context context;
    private Paint paint;
    private Bitmap bmp;
    public GameSurfaceView(Context context)
    {
        super(context);
        try {
            SurfaceHolder holder = getHolder();
            holder.addCallback(this);
            this.context = context;
        }catch (Exception e)
        {
            e.printStackTrace();
        }
    }

    public void render(Canvas canvas)
    {
        canvas.drawBitmap(bmp,0,0,paint);
    }
    @Override
    public void surfaceCreated(SurfaceHolder holder) {
        bmp=BitmapFactory.decodeResource(context.getResources(),
                                                  R.drawable.img);
        paint = new Paint();
        renderThread = new RenderThread(this);
        renderThread.start();
    }

    @Override
    public void surfaceChanged(SurfaceHolder holder, int format, int
                              width, int height) {
    }

    @Override
    public void surfaceDestroyed(SurfaceHolder holder) {
        renderThread.end();
    }
}
```

在 surfaceCreate 中加载资源、创建画笔和启动渲染线程，surfaceDestroyed 结束渲染线程。渲染线程的代码如下：

```
public class RenderThread extends Thread{
    private GameSurfaceView view;
    private SurfaceHolder holder;
    private boolean bRun = true;
    public RenderThread(GameSurfaceView view){
        this.view = view;
        holder = view.getHolder();
    }

    @Override
    public void run()
    {
        long startTime = 0;
        long endTime = 0;
        int minLoopTime = 1000/60;

        Canvas c;
        while(bRun)
        {
            startTime = System.currentTimeMillis();
            c=null;
            try{
                c=holder.lockCanvas();
                synchronized (holder)
                {
                    view.render(c);
                }
            }finally {
                if(c!=null)
                {
                    holder.unlockCanvasAndPost(c);
                }
            }

            endTime = System.currentTimeMillis();
            long sleepTime = minLoopTime - (endTime - startTime);
            if(sleepTime>0)
            {
```

```
                try {
                    Thread.sleep(sleepTime);
                }catch (InterruptedException e)
                {
                    e.printStackTrace();
                }
            }
        }
    }

    public void end()
    {
        bRun = false;
    }
}
```

在线程的运行代码每次循环加入帧频控制代码，每帧开始时计时，当帧结束时查看经过的时间，如果少于指定时间，则通过调用 sleep( )方法睡眠所剩时间，从而控制帧频。

7 OpenGL ES 图形渲染

SurfaceView 对于一般的 2D 游戏开发已经足够使用，如果要开发 3D 游戏，则需要使用 OpenGL ES。OpenGL 是一个专业的 3D 程序接口，一个功能强大、调用方便的底层 3D 图形库。Android 3D 引擎采用的是 OpenGL ES。它是基于 OpenGL 的子集，是一套专门为手持和嵌入式系统设计的 3D 引擎 API。使用 OpenGL ES 开发 3D 游戏会涉及大量图形学内容。限于篇幅，本章简要阐述 OpenGL1.0 固定渲染管线，以及使用 OpenGL 开发二维游戏会涉及的重要概念，更多内容请参照其他书籍。

7.1 GLSurfaceView

GLSurfaceView 是 Android 应用程序中实现 OpenGL 画图的重要组成部分。SurfaceView 是 View 的子类，而 GLSurfaceView 则是 SurfaceView 的子类。GLSurfaceView 需要通过 GLSurfaceView::Renderer(渲染器)接口来进行渲染。Renderer 是 GLSurfaceView 的内部静态接口。

此接口定义了一个 OpenGL 在 GLSurfaceView 上作画所需的方法。开发时需要自定义一个类来实现这个接口，然后设置为 GLSurfaceView 的渲染器。GLSurfaceView 和 Renderer 的关系类似画布和画笔。GLSurfaceView 在继承 SurfaceView 的基础上，自行创建了渲染线程，并在线程中调用 Renderer 的相关方法，Renderer 接口定义了下面三个方法：

- onSurfaceCreated( )　当创建 GLSurfaceView 时被调用，只调用一次。在这个方法中执行只发生一次的动作，比如设置 OpenGL 环境参数或初始化 OpenGL 图形对象。
- onDrawFrame( )　系统在每次重绘 GLSurfaceView 时调用此方法，此方法是绘制图形对象的主要的执行点。
- onSurfaceChanged( )　当 GLSurfaceView 几何体改变时系统调用此方法，比如 GLSurfaceView 的大小改变或设备屏幕的方向改变。使用此方法来响应 GLSurfaceView 容器的变化。

可以发现 Renderer 的三个方法与 SurfaceHolder. Callback 的三个方法相近。实际上，GLSurfaceView 封装时分别在 Callback 的 SurfaceCreate 上调用了 onSurfaceCreated，当然并非直接调用，而是通过封装在内部线程完成的。而 surfaceChanged 间接调用了 onSurfaceChanged，surfaceDestoryed 在 Renderer 中没有定义相应方法，而是自行做了线程的结束处理。onDrawFrame 对应前面 SurfaceView 中自定义的 Renderer 方法，需要在其

中写入帧的渲染代码。下面是使用 GLSurfaceView 渲染的代码：


```
public class GameGLView extends GLSurfaceView {
    private GameRenderer renderer;
    public GameGLView(Context context) {
        super(context);
        renderer = new GameRenderer();
        setRenderer(renderer);
    }

    private class GameRenderer implements GLSurfaceView.Renderer
    {

        @Override
        public void onSurfaceCreated(GL10 gl, EGLConfig config) {
            gl.glClearColor(1.0f, 0.0f, 0.0f, 0.5f);
            gl.glShadeModel(GL10.GL_SMOOTH);
            gl.glClearDepthf(1.0f);
            gl.glEnable(GL10.GL_DEPTH_TEST);
            gl.glDepthFunc(GL10.GL_LEQUAL);
            gl.glHint(GL10.GL_PERSPECTIVE_CORRECTION_HINT,GL10.
                    GL_NICEST);
        }

        @Override
        public void onSurfaceChanged(GL10 gl, int width, int height) {
            gl.glViewport(0,0,width,height);
        }

        @Override
        public void onDrawFrame(GL10 gl) {
            gl.glClear(GL10.GL_DEPTH_BUFFER_BIT | GL10.GL_COLOR
                    _BUFFER_BIT);

        }
    }
}
```

虽然上述代码只是设置了背景色，没渲染任何对象，但代码比 SurfaceView 更简洁，因为 SurfaceHolder 和渲染线程已经封装到 GLSurfaceView 内部。实现 Renderer 的 GameRenderer 类也放到 GameGLView 中作为内部类处理。

7.2 渲染管线

渲染管线(graphics pipeline)指 OpenGL 在渲染处理过程中执行的一系列操作，或者说渲染的处理流程，管线的各个阶段实际上是由 GPU 的图形处理单元并行处理，这样可以极大地提高渲染效率。渲染管线在 1.x 的阶段是采用固定渲染管线，如图 7 - 1 所示。

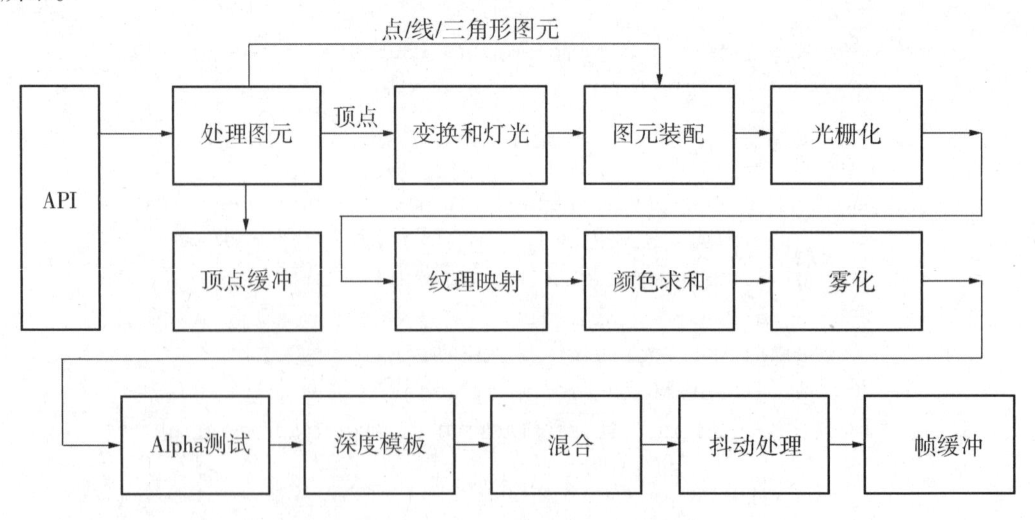

图 7 - 1　OpenGL1.1 的渲染管线示意图

从图 7 - 1 可以看出，渲染管线分成很多个阶段：①对顶点数据做基本处理；②进行坐标变换和光照计算；③选定图元，进行光栅化处理，将几何图形转换成二维图像；④进行纹理映射，计算像素颜色；⑤进行雾处理；⑥做一系列测试(Alpha、深度、模板)以及混合和抖动计算处理；⑦写入帧缓冲。

7.3 顶点和图元

OpenGL ES 支持绘制的基本几何图形分为三类：点、线段、三角形。任何复杂的 2D 或 3D 图形都是通过这三种几何图形构造而成的。而无论是点、线段还是三角形都是通过顶点来定义的。顶点包括的属性比较多，根据渲染目标不同，可能会用到顶点空间坐标、纹理坐标、颜色、法线等。顶点采用顶点数组存放，将不同顶点属性值传给渲染管线。下面是正方形类的代码：

```
public class Square {
     FloatBuffer vertexBuffer;
     FloatBuffer colorBuffer;

    public Square()
    {
        initVertices();
    }

    public void initVertices()
    {
        float[] vertices = new float[]
                {
                        -0.5f,0.5f,0.0f,
                        -0.5f, -0.5f,0.0f,
                        0.5f,0.5f,0.0f,
                        0.5f, -0.5f,0.0f,
                };
        vertexBuffer = bufferUtil(vertices);

        float[] colors = new float[]
                {
                            0.0f,0.0f,1.0f,1.0f,
                            0.0f,1.0f,0.0f,1.0f,
                            1.0f,0.0f,0.0f,1.0f,
                            1.0f,1.0f,0.0f,1.0f
                };
        colorBuffer = bufferUtil(colors);

    }

    public void render(GL10 gl)
    {
        gl.glEnableClientState(GL10.GL_VERTEX_ARRAY);
        gl.glEnableClientState(GL10.GL_COLOR_ARRAY);

        gl.glVertexPointer(3, GL10.GL_FLOAT, 0, vertexBuffer);
        gl.glColorPointer(4, GL10.GL_FLOAT, 0, colorBuffer);

        gl.glDrawArrays(GL10.GL_TRIANGLE_STRIP, 0,4);

        gl.glDisableClientState(GL10.GL_COLOR_ARRAY);
        gl.glDisableClientState(GL10.GL_VERTEX_ARRAY);
    }
```

```
    public FloatBuffer bufferUtil(float []arr)
    {
        FloatBuffer buffer;
        ByteBuffer vbb = ByteBuffer.allocateDirect(arr.length* 4);
        vbb.order(ByteOrder.nativeOrder());
        buffer = vbb.asFloatBuffer();
        buffer.put(arr);
        buffer.position(0);
        return buffer;
    }
}
```

在上述代码中，initVertices 方法用于通过顶点构建一个正方形，其中定义了两个浮点型数组，vertices 存放空间坐标，而 colors 存放顶点颜色。将两个浮点数组转换成 FloatBuffer 之后可以指定给渲染管线，分别用于读取坐标和颜色数据。通过 glDrawArrays 来渲染顶点，方法中第一个参数表示采用的图元类型，第二个参数表示渲染采用的起始顶点在数组中下标，第三个参数表示渲染用的顶点数量。在 GL10 中，图元类型有七种，如表 7－1 所示。

表 7－1　GL10 的图元类型

图元名	说　明
GL_POINTS	点列，逐个孤立渲染的顶点
GL_LINES	线列，每两个顶点组成一条线段
GL_LINE_STRIP	线带，每两个相邻的顶点都连成线段
GL_LINE_LOOP	闭合线段，除相邻顶点外，首尾两个顶点也相连
GL_TRIANGLES	三角形列，每三个顶点组成一个三角形
GL_TRIANGLE_STRIP	三角形带，每三个相邻顶点组成三角形
GL_TRIANGLE_FAN	三角扇形，首顶点和每对相邻两个顶点构成三角形

具体连接的几何形状如图 7－2 所示。

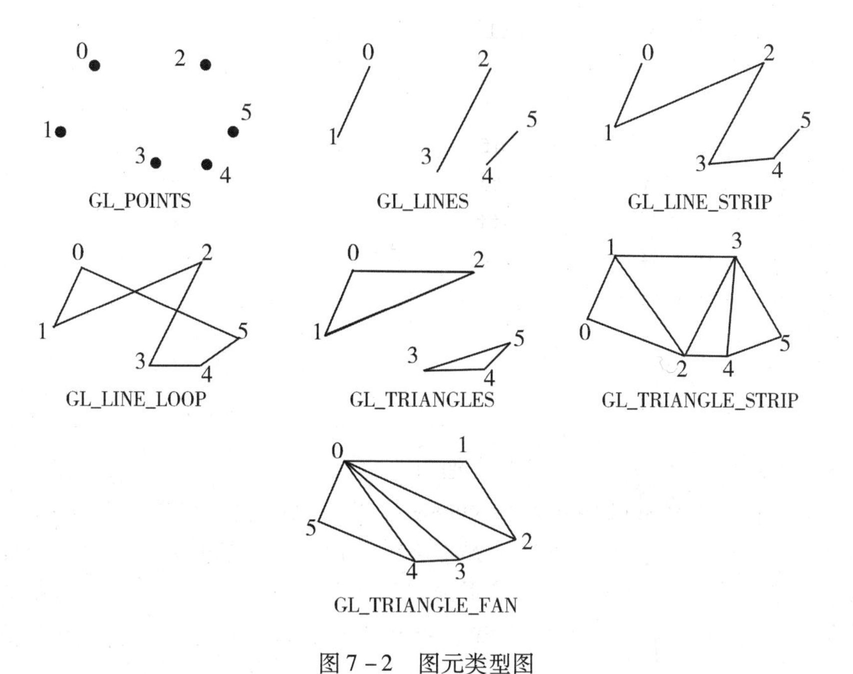

图 7－2　图元类型图

在创建正方形 Square 类之后，可以渲染正方形，在 GameGLView 中创建 Square 实例，并在 onDrawFrame 中调用其 render 方法，代码如下：

```
public class GameGLView extends GLSurfaceView {
    private GameRenderer renderer;
    Square square;
    public GameGLView(Context context) {
        super(context);
        renderer = new GameRenderer();
        setRenderer(renderer);
        square = new Square();
    }

    private class GameRenderer implements GLSurfaceView.Renderer
    {
    @Override
    public void onSurfaceCreated(GL10 gl, EGLConfig config) {
        gl.glClearColor(1.0f, 0.0f, 0.0f, 0.5f);
        gl.glShadeModel(GL10.GL_SMOOTH);
        gl.glClearDepthf(1.0f);
```

```
            gl.glEnable(GL10.GL_DEPTH_TEST);
            gl.glDepthFunc(GL10.GL_LEQUAL);
            gl.glHint(GL10.GL_PERSPECTIVE_CORRECTION_HINT,GL10.GL_NICEST);
        }

        @Override
        public void onSurfaceChanged(GL10 gl, int width, int height) {
            gl.glViewport(0, 0, width, height);
        }

        @Override
        public void onDrawFrame(GL10 gl) {
            gl.glClear(GL10.GL_DEPTH_BUFFER_BIT | GL10.GL_COLOR_BUFFER_BIT);
            square.render(gl);
        }
    }
}
```

运行之后效果如图7-3所示。效果看起来并不是一个正方形，这是因为目前并未设置正确的变换，所以正方形被拉伸变形。

图7-3 项目运行效果图

7.4 坐标变换

建模时设定的坐标值需要经过世界、观察和投影三步变换才能正确转换到显示屏幕上，变换通过矩阵进行。这一点在第6章位图变换中已经提到，但在3D图形变换采用的矩阵与之前不同，之前矩阵是3×3矩阵(android. graphics. Matrix)，而随着维度升高到3维，就需要采用4×4矩阵(android. opengl. Matrix)。

7.4.1 观察变换

随着摄像机(camera)选择的位置、角度和方向的不同，屏幕上渲染的画面也不同，通过观察变换来确定摄像机的拍摄方式。OpenGL的GLU中提供了一个方法来帮助计算观察矩阵。

```
public static void gluLookAt(GL10 gl, float eyeX, float eyeY, float eyeZ,
        float centerX, float centerY, float centerZ, float upX, float upY,
        float upZ)
```

eyeX、eyeY、eyeZ分别表示摄像机所处位置的三维坐标，centerX、centerY、centerZ分别表示观察目标点的三维坐标，而upX、upY、upZ则表示摄像机顶端指向的方向。通过设置这些参数，可以确定摄像机在场景中所有的属性。设置摄像机代码如下：

```
gl.glMatrixMode(GL10.GL_MODELVIEW);
gl.glLoadIdentity();
GLU.gluLookAt(gl, 0, 0, 5, 0f, 0f, 0f, 0f, 1.0f, 0.0f);
```

这段代码表示将摄像机放置在(0，0，5)的位置，向(0，0，0)目标点观察，摄像机顶端为(0，1，0)，表示竖直向上。glMatrixMode表示选择矩阵模式，有三个模式可选，即ModelView(模型视图)、Projection(投影)、Texture(纹理)，这三种模式对应OpenGl管理的三个矩阵堆栈。堆栈的大小容量根据需求不同而改变。例如ModelView需要管理世界和观察变换矩阵，相应变化较多，最多可容纳16个矩阵，而投影矩阵和纹理矩阵通常变化少，能容纳2个矩阵。

当需求操作某种矩阵之前，需要切换到相应的矩阵模式，上述代码中，需要设置观察矩阵，所以要设置ModelView模式。同样下节要设置投影矩阵前，相应切换到投影模式。OpenGL使用堆栈管理，提供了两个重要的方法glPushMatrix和glPopMatrix，分别用于入栈和出栈操作，当glPushMatrix时，栈顶的矩阵会复制一份入栈，这时调用的矩阵操作函数改变的就是原本栈顶矩阵的一个拷贝。当使用完之后，只需调用glPopMatrix将拷贝的矩阵出栈，就恢复原本栈的数据。

在上述代码中，设置了相应的模型视图矩阵模式后，为了防止之前栈顶矩阵的数据干扰，调用glLoadIDentity()先将当前矩阵设为单位阵，在此基础上再调用gluLookAt方法计算观察矩阵。

7.4.2 投影变换

确定摄像机之后，需要把观察的3D世界的场景投影到二维平面上，这个过程通过投影变换来完成。投影变换分为正交投影和透视投影。两者都是建立视景体，裁剪视景体之外的部分，视景体内的部分通过投影将三维坐标转到二维坐标。两种投影变换的区别在于建立的视景体不同。透视投影的视景体如图7-4所示。

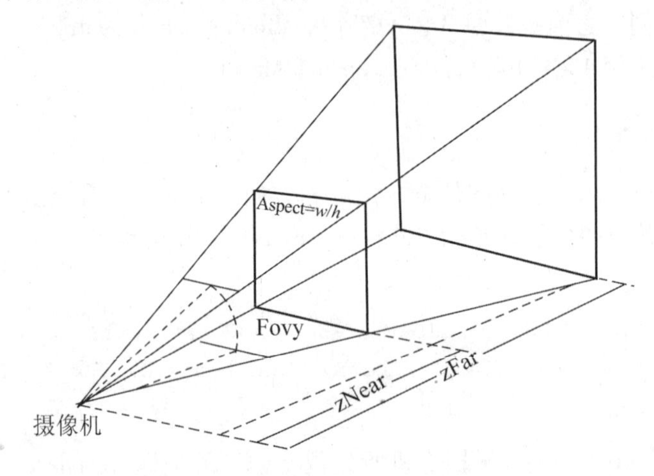

图7-4　gluPerspective 透视投影

如图7-4所示，从摄像机出发，指定其视角Fovy创建一个棱锥，设定近截切平面zNear是摄像机与之的距离、远截切平面zFar是摄像机与之的距离，设定纵横比，从而得到视景体，通过gluPerspective来计算投影矩阵。

```
void gluPerspective(GLdouble fovy,GLdouble aspect,GLdouble zNear, GLdouble
zFar);
```

另一个方法如图7-5所示，设定近截切平面的左(left)、右(right)、上（top)、下(bottom)，以及和远截切平面的远近值Near和Far来创建视景体，通过方法glFrustum计算投影矩阵。

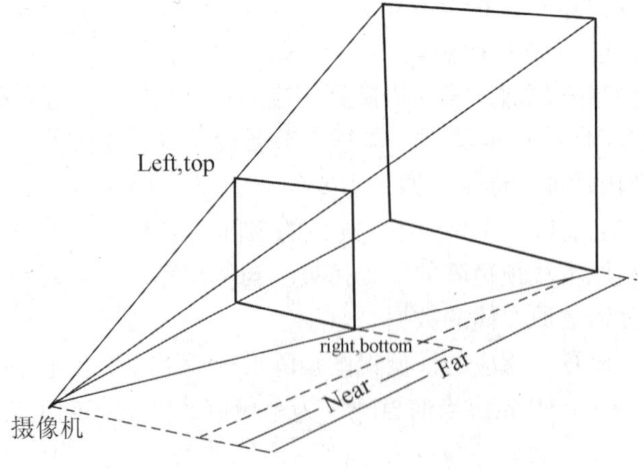

图7-5　glFrustum 投影矩阵

```
void glFrustum(GLdouble left,GLdouble Right,GLdouble bottom,GLdouble top,
GLdouble near,GLdouble far);
```

透视投影下，可视对象与现实世界人们的视觉效果一致。近大远小、三维游戏中的对象绝大部分采用透视投影。

正交投影创建的是平行视景体，也就是长方体，如图7-6所示。

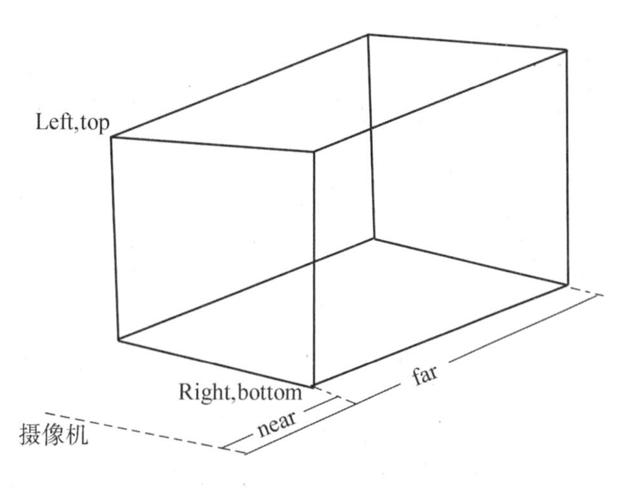

图7-6 正交投影

正交投影远近截切平面同样大小，中间的对象也采用平行投影。无论远近、大小保持不变。正交投影通常用于二维的图形渲染，如游戏中的UI界面或某些特效，或者直接在二维游戏的开发中用到。同样有两个不同的方法计算正交投影矩阵。

```
glOrtho (GLdouble left, GLdouble right, GLdouble bottom, GLdouble top,
GLdouble near, GLdouble far)
gluOrtho2D(left, right, bottom, top)
```

这两个方法其实差不多，后者是对前者的封装，其中最后两个参数分别设置为-1.0f和1.0f。给渲染加上观察和投影变换之后代码如下：

```
public void onSurfaceCreated(GL10 gl, EGLConfig config) {
      gl.glClearColor(1.0f, 0.0f, 0.0f, 0.5f);
      gl.glShadeModel(GL10.GL_SMOOTH);
      gl.glClearDepthf(1.0f);
      gl.glEnable(GL10.GL_DEPTH_TEST);
      gl.glDepthFunc(GL10.GL_LEQUAL);
      gl.glHint(GL10.GL_PERSPECTIVE_CORRECTION_HINT,GL10.GL_NICEST);
}

@ Override
public void onSurfaceChanged(GL10 gl, int width, int height) {
```

```
        gl.glViewport(0, 0, width, height);

        float ratio = (float) width / height;
        gl.glMatrixMode(GL10.GL_PROJECTION);
        gl.glLoadIdentity();
        gl.glFrustumf(-ratio, ratio, -1, 1, 1, 10);
    }

    @Override
    public void onDrawFrame(GL10 gl) {
        gl.glClear(GL10.GL_DEPTH_BUFFER_BIT | GL10.GL_COLOR_BUFFER_BIT);
        gl.glMatrixMode(GL10.GL_MODELVIEW);
        gl.glLoadIdentity();
        GLU.gluLookAt(gl, 0, 0, 5, 0f, 0f, 0f, 0f, 1.0f, 0.0f);
        square.render(gl);
    }
```

7.4.3 模型变换

对于游戏中的物体，绝大部分使用的投影矩阵和观察矩阵是一样的，但都有各自的模型变换矩阵。模型变换通过平移、旋转、缩放等将模型从模型空间变换到世界空间，这个过程实际上就是构建3D虚拟世界的过程。每个物体通过模型变换设定在世界中的位置、角度及大小，组成整个世界。通过基本矩阵的相乘得到复合矩阵，这与6.1.2节中位图使用的矩阵规则一样，不同之处在于OpenGL中使用的GL矩阵操作方法大部分是生成矩阵与当前矩阵栈顶的矩阵进行设置，而根据其执行顺序，都是依次右乘，所以在OpenGL代码中，要注意的是后执行得到的矩阵反而会在最终复合矩阵中先运行。例如，代码

```
gl.glRotatef(45.0f,0.0f,0.0f,1.0f);
gl.glTranslatef(5.0f,0.0f,0.0f);
```

得到的综合矩阵即旋转矩阵×平移矩阵，效果则是先平移再旋转。

7.5 纹理

无论是二维游戏还是三维游戏的开发，都需要用到位图资源，OpenGL显示位图的方式是通过纹理映射。

7.5.1 纹理坐标

OpenGL加载位图资源来创建纹理，渲染管线将纹理准确地映射到相应的几何图元

上，是通过纹理坐标来指定的，通过给每个顶点设置纹理坐标确定映射的过程。纹理坐标与位置坐标不同，它是一个二维坐标，取纹理图的左上角作为坐标原点，两条坐标轴 u 横向右，v 纵向下，整个纹理图宽为 u 的一个单位长度，高为 v 的一个单位长度，如图 7－6 所示。

有了纹理坐标系，可以确定纹理中任意纹理元的位置，甚至在纹理图之外（u、v 值在[0，1]范围之外）也可以表示。只要为模型每个顶点指定相应的纹理坐标，就可以确定纹理映射方式。

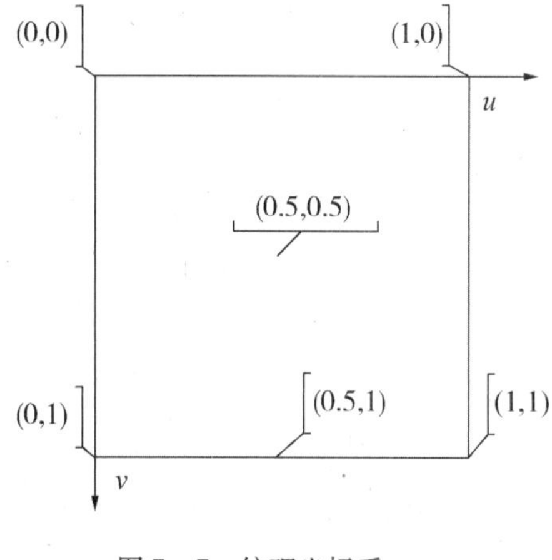

图 7－7　纹理坐标系

7.5.2　创建纹理

创建纹理需要给纹理分配一个编号，这个编号通过方法 glGenTexture 获得，得到编号之后再将位图资源与纹理绑定。代码如下：

```
Bitmap bmp = BitmapFactory.decodeResource (context.getResources ( ),
R.drawable.img);
int[] textures = new int[1];
IntBuffer intBuffer = IntBuffer.allocate(1);
//生成纹理编号
gl.glGenTextures(1,intBuffer);
//得到当前纹理编号
int currTextureid = intBuffer.get();//textures[0];
//绑定到纹理
gl.glBindTexture(GL10.GL_TEXTURE_2D, currTextureid);
//将位图资源与纹理绑定
GLUtils.texImage2D(GL10.GL_TEXTURE_2D, 0, bmp, 0);
```

纹理创建成功，渲染时就通过纹理编号来指定使用的纹理。

7.5.3　纹理映射

要将一张位图直接映射到 7.4 节渲染的正方形中，首先需要确定四个顶点各自的纹理坐标。如果无需剪切，那么四个顶点对应的纹理坐标分别设置如下：

```
float[] texCoords = new float[]{0.0f,0.0f, 0.0f,1.0f, 1.0f,0.0f, 1.0f,
1.0f};
```

由于位图的尺寸未必是正方形，因此需要将正方形进行拉伸，通过缩放矩阵可以完成。横方向缩放系数由位图宽度决定，纵方向缩放系数由位图高度决定。更改后的 Square 代码如下：

```
public class Square {

    FloatBuffer vertexBuffer;
    FloatBuffer colorBuffer;
    FloatBuffer texCoordBuffer;

    int texid;

    public Square(int texid)
    {
        initVertices();
        this.texid = texid;
    }

    public void initVertices()
    {
        float[] vertices = new float[]
                {
                        -0.5f,0.5f,0.0f,
                        -0.5f, -0.5f,0.0f,
                        0.5f,0.5f,0.0f,
                        0.5f, -0.5f,0.0f
                };
        vertexBuffer = bufferUtil(vertices);

        float[] colors = new float[]
                {
                        0.0f,0.0f,1.0f,1.0f,
                        0.0f,1.0f,0.0f,1.0f,
                        1.0f,0.0f,0.0f,1.0f,
                        1.0f,1.0f,0.0f,1.0f
                };
        colorBuffer = bufferUtil(colors);

        float[] texCoords = new float[]
                {
                        0.0f,0.0f,
                        0.0f,1.0f,
                        1.0f,0.0f,
                        1.0f,1.0f
                };
```

```
        texCoordBuffer = bufferUtil(texCoords);
    }

    public void render(GL10 gl)
    {
        gl.glEnableClientState(GL10.GL_VERTEX_ARRAY);
        gl.glEnableClientState(GL10.GL_COLOR_ARRAY);
        gl.glEnableClientState(GL10.GL_TEXTURE_COORD_ARRAY);
        gl.glEnable(GL10.GL_TEXTURE_2D);

        gl.glVertexPointer(3, GL10.GL_FLOAT, 0, vertexBuffer);
        gl.glColorPointer(4, GL10.GL_FLOAT, 0, colorBuffer);
        gl.glTexCoordPointer(2,GL10.GL_FLOAT,0,texCoordBuffer);
        gl.glBindTexture(GL10.GL_TEXTURE_2D, texid);

        gl.glDrawArrays(GL10.GL_TRIANGLE_STRIP, 0, 4);

        gl.glDisable(GL10.GL_TEXTURE_2D);
        gl.glDisableClientState(GL10.GL_TEXTURE_COORD_ARRAY);
        gl.glDisableClientState(GL10.GL_COLOR_ARRAY);
        gl.glDisableClientState(GL10.GL_VERTEX_ARRAY);
    }

    public FloatBuffer bufferUtil(float []arr)
    {
        FloatBuffer buffer;
        ByteBuffer vbb = ByteBuffer.allocateDirect(arr.length* 4);
        vbb.order(ByteOrder.nativeOrder());
        buffer = vbb.asFloatBuffer();
        buffer.put(arr);
        buffer.position(0);
        return buffer;
    }
}
```

8 数据存储访问

在 Android 平台，数据存储可以通过四种方式来实现：

(1)SQLite。SQLite 是一个轻量级的数据库，支持基本 SQL 语法，是移动平台经常采用的一种数据存储方式。Android 为此数据库提供了一个名为 SQLiteDatabase 的类，封装了一些操作数据库的 API。

(2)SharedPreference。这是除 SQLite 数据库外，另一种常用的数据存储方式。其本质就是一个 XML 文件，常用于存储较简单的参数设置。

(3)File。文件系统，即文件(I/O)存储方法，常用于存储大数量的数据，缺点是更新数据比较困难。

(4)ContentProvider。这是 Android 系统中能实现所有应用程序共享的一种数据存储方式。因为数据通常在各应用间是彼此私密的，所以此存储方式较少使用，但是它又是必不可少的一种存储方式。例如，音频、视频、图片和通讯录一般都可以采用这种方式进行存储。

8.1 本地存储

Android 平台提供了一个 SharedPreferences 类，它是一个轻量级的存储类，可以用于数据的存储，但该类一般用于保存应用程序的参数配置等小型数据。该类实际上是通过 XML 文件来实现，文件存放在/data/data/ < package name >/shared_prefs 目录下。也就是说，数据是通过 name - value 对的方式存储数据，可以存储一些基本数据类型，包括 boolean、integer、float、string 等。

8.1.1 SharedPreferences 写入

实现 SharedPreferences 存储过程并不复杂，完整步骤如下：

(1)根据 Context 获取 SharedPreferences 对象。

(2)利用 edit()方法获取 Editor 对象。

(3)通过 Editor 对象存储 key - value 键值对数据。

(4)通过 commit()方法提交数据。

首先要获得 SharedPreferences 对象，通过 Context 类的 getSharedPreferences 方法获得一个 SharedPreferences 对象。如果在 Activity 类的方法中，可以直接调用，否则要先获得当前应用的 context 引用来调用该方法(Activity 的自身引用 this 就是该对象)。在 Activity

的方法中创建 SharedPreferences 的代码如下：

```
SharedPreferences sp = getSharedPreferences("SP", MODE_PRIVATE);
```

getSharedPreferences(name, mode)方法的第一个参数用于指定配置文件的名称，名称不用带后缀，后缀会由 Android 访问自动加上。方法的第二个参数指定文件的访问方式，访问方式有四种，如表 8－1 所示。

表 8－1　文件打开方式

方　式	说　明
MODE_APPEND	在文件尾写入数据(追加)方式
MODE_PRIVATE	只可应用程序自身读写模式
MODE_WORLD_READABLE	其他应用程序对文件可读模式
MODE_WORLD_WRITEABLE	其他应用程序对文件可写模式

由于 SharedPreferences 对象本身只能获取数据而不支持存储和修改，存储修改是通过 Editor 对象实现的，通过 Editor 对象来进行写入和提交。代码如下：

```
Editor editor = sp.edit();
editor.putString("STRING_KEY", "string");
editor.putBoolean("BOOLEAN_KEY", true);
editor.commit();
```

写入完成之后，可以在/data/data/ <package name>/shared_prefs 目录下找到相应的 sp.xml 文件，其内容如下：

```
<?xml version='1.0' encoding='utf-8' standalone='yes' ?>
<map>
        <string name="STRING_KEY">string</string>
        <boolean name="BOOLEAN_KEY" value="true" />
</map>
```

8.1.2　SharedPreferences 读取

读取比写入简单，无需通过 Editor 对象进行。在获取指定的 SharedPreferences 对象后，根据相应 name 读取其 value 数据。

```
String strValue = sp.getString("STRING_KEY", "none"));
```

此处 getString 同样有两个参数，第一个是数据对中的 name 值，通过其在 XML 中找到对应数据值，第二个参数则是没有 name 对应的值时返回的默认值。在上一句代码中，如果没有 STRING_KEY 对应的数据，则 strValue 被赋予“none”字符串。

8.2 文件存储

SharedPreferences 只能存储 boolean、int、float、long 和 String 五种简单的数据类型，无法进行条件查询等。一般用于应用程序的简单参数配置。如果是比较大型的数据，可以采用文件的方式来存储。Android 平台基于 Linux 系统，对文件访问需要具备用户权限，应用允许访问的文件可以是在 SD 卡或应用自身目录下(/data/data/ < package_ name >/files)。

在文件 I/O 操作上，除了标准的 Java I/O 类和方法外，设备环境 Context 类提供了一些方法用于文件的读/写操作以及目录的管理，如表 8 -2 所示。

表 8 -2 Android 文件操作方法

方 法	说 明
openFileInput	用于读取当前应用文件夹下的文件，并返回 FileInputStream(输入流)
openFileOutput	用于向当前应用文件夹下输出文件，并返回 FileOutputStream(输出流)
deleteFile	删除当前应用文件夹下文件
fileList	获得当前应用文件夹下文件列表
getFilesDir	获得当前应用文件夹下目录，即/data/data/ < package_ name >/files
getCacheDir	获得当前应用缓存目录，为/data/data/ < package_ name >/cache
getDir	获取或创建应用文件子目录

表 8 -2 中的方法都只支持操作当前 Android 应用程序文件夹下的文件，即应用的私有目录，并且传入的文件名参数不能带有任何路径信息，只需要传入文件名即可(包括扩展名)。当创建文件时，如果指定的文件不存在，则 Android 会创建文件，而对于存在的文件，默认使用覆盖私有模式(Context. MODE_PRIVATE)对文件进行写操作。如果想让增量方式写入已存在的文件，需要指定输出模式为 Context. MODE_ APPEND，和前面 SharedPreferences 打开文件的方式一样。如果打算允许让其他应用访问输出的文件，可以设置输出模式为只读(Context. MODE_ WORLD_ READABLE)或可读写(Context. MODE_ WORLD_ WRITEABLE)。

8.2.1 应用文件目录访问

通过 openFileInput 获得文件输入流，然后通过输入流读取文件数据到缓冲，将其解析成字符串。

```
try{
FileInputStream fis = openFileInput(GameConst.Datastorage.Game_SAVE);
byte[] buff = new byte[1024];
int hasRead = 0;
StringBuilder fileStr = new StringBuilder("");
while((hasRead = fis.read(buff))>0){
fileStr.append(EncodingUtils.getString(buff,"utf-8"));
}
Log.i(TAG,fileStr.toString());
}catch(Exception e){
e.printStackTrance();
}
```

反之要写入文件，则通过 openFileOutput 获得文件输出流，然后通过输出流写入文件数据到缓冲，将其解析成字符串。

```
try{
FileOutputStream fos = openFileOutput(GameConst.Datastorage.Game_
                                        SAVE,MODE_PRIVATE);
String dataStr = "testStr";
fos.write(dataStr.getBytes("utf-8"));
fos.close();
}catch(Exception e){
e.printStackTrace();
}
```

在调用 openFileOutput 时，它的第二个参数即为文件打开方式，此处 MODE_PRIVATE 为覆盖写入。

8.2.2 SD 卡访问

文件除了可以存放在应用自身的应用文件夹下，还可以放在扩展存储设备 SD 卡中。要访问 SD 卡，首先应用需要向系统申请获得访问 SD 卡的权限，在项目的描述文件 AndroidManifest.xml 中加入需求权限的代码，分别是创建删除文件权限和写入文件权限。

```
<?xml version="1.0" encoding="utf-8"?>
<manifest xmlns:android="http://schemas.android.com/apk/res/android"
     package="com.jeremy.helloworld" >
   <!-- SDCard 中创建与删除文件权限 -->
       <uses-permission android:name="android.permission.MOUNT_UNMOUNT_
FILESYSTEMS"/>
       <!-- 向 SDCard 写入数据权限 -->
       <uses-permission android:name="android.permission.WRITE_EXTERNAL
_STORAGE"/>
```

```
    <application
        android:allowBackup = "true"
        android:icon = "@ mipmap/ic_launcher"
        android:label = "@ string/app_name"
        android:supportsRtl = "true"
        android:theme = "@ style/AppTheme" >
        <activity android:name = ". MainActivity" >
            <intent - filter >
                <action android:name = "android. intent.
                 action. MAIN" / >
                <category android:name = "android. intent.
                 category. LAUNCHER"/ >
            </intent - filter >
        </activity >
    </application >
</manifest >
```

在最新推出的Android6.0系统中，权限机制有了变化，Google将权限分为普通权限和危险权限两种，访问网络之类属于普通权限，像上述代码在manifest中申请即可；但SD卡访问就属于危险权限，需要用户进行授权。目前，我们开发的应用运行于Android6.0以下版本，按照上面这种方式即可。

申请权限之后，我们需要获取SD卡的相关信息，操作相关方法如表8-3所示。

表8-3　SD卡访问方法

方　法	说　明
getDataDirectory	获取SD卡中data目录路径
getDownloadCacheDirectory	获取SD卡中download目录路径
getExternalStorageDirectory	获取SD卡路径
getExternalStorageState	获取SD卡状态
getRootDirectory	获取Root路径

访问SD卡，首先要确认SD卡存在并且正常，这个通过调用表中查询状态语句来进行。

```
if (Environment. getExternalStorageState (). equals (Environment. MEDIA_MOUNTED)) {
}
```

SD卡的状态有很多种，包括不存在、处于SD卡USB共享以及只读等，读者可以结合自己使用Android平台的经验来对照，其中只有MEDIA_MOUNTED正常挂载状态可以进行SD卡的读写操作。如果状态正常，则可以获取SD卡路径来访问其中的文件。

读取文件数据代码如下：

```
try {
if (Environment. getExternalStorageState (). equals (Environment. MEDIA_MOUNTED)){
    FileInputStream fis = new FileInputStream (
          Environment. getExternalStorageDirectory () + "/" +
          GameConst. Datastorage. Game_SAVE);

  byte[] buff = new byte[1024];
  int hasRead = 0;
StringBuilder fileStr = new StringBuilder ("");
while ((hasRead = fis. read (buff)) >0){
fileStr. append (EncodingUtils. getString (buff, "utf -8"));
}
Log. i (TAG, fileStr. toString ());
}
}catch (Exception e){
e. printStackTrance ();
}
```

写入文件代码如下：

```
try{
if (Environment. getExternalStorageState (). equals (Environment. MEDIA_MOUNTED)){
FileOutputStream fos =
openFileOutput (Environment. getExternalStorageDirectory () + "/" + GameConst.
Datastorage. Game_SAVE, MODE_PRIVATE);
String dataStr = "testStr";
fos. write (dataStr. getBytes ("utf -8"));
fos. close ();
}
}catch (Exception e){
e. printStackTrace ();
}
```

使用文件存储相比 SharedPreferences 可以存放大些的数据，但是通过字节的方式写入文件，存储和读取时要做相应的数据解析，并不适合大量不同类型数据存放，适合存放一些文本数据或者音乐图片等。

8.3 SQLite

SQLite 是开源的一个嵌入式关系数据库，与一般关系型数据库相比，它比较简单、高效、占用资源少，所以在移动平台上应用很广泛。Android 为了更好地访问 SQLite，还提供了一个 SQLiteOpenHelper 类，帮助我们访问 SQLite 数据库。

8.3.1 SQLiteOpenHelper 类

SQLiteOpenHelper 是个抽象类，不能直接使用，需要创建一个新类来继承它。其中有两个抽象方法需要在子类重写，这两个方法分别是 onCreate 和 onUpgrade，从命名上可以看出，分别用于创建和更新升级数据库。下面代码创建了一个子类。

```
public class GameDBHelper extends SQLiteOpenHelper {
private static final String TAG ="GameDBHelper";

public GameDBHelper(Context context, String name) {
super(context, name, null, 1);
}

@ Override
public void onCreate(SQLiteDatabase db) {
try{
StringBuffer initSql = new StringBuffer();
initSql.append("create table");
initSql.append(GameConst.DataStorage.TABLE_NAME);
initSql.append("(");
initSql.append("UserName");
initSql.append(" text,");
initSql.append("UserLevel");
initSql.append(" integer);");
db.execSQL(initSql.toString());
}catch(Exception e){
e.printStackTrace();
}
}

@ Override
public void onUpgrade(SQLiteDatabase db, int oldVersion, int newVersion) {
}
}
```

上述代码定义了一个 SQLiteOpenHelper 的子类，并在其中重写了 onCreate 和 onUpgrade 方法，onCreate 方法在数据库创建时被调用。在这里创建了一个表，并定义了其中两个字段。OnUpgrade 则是在数据库版本发生变化时调用，版本号在构造函数时指定，通常数据库表的结构发生改变时会修改版本号，在此处可能需要将原本数据读出来，写入新表。

8.3.2 SQLite 插入删除及修改

实例化 GameDBHelper 对象来操作 SQLite 数据库。要访问 SQLite 数据库中的数据，需要得到 SQLite 数据库对象 SQLiteDatabase。该对象提供了 execSQL 方法用于执行 SQL 语句，如在 7.3.1 节的 onCreate 方法中创建表。onCreate 方法已传入了 SQLiteDatabase 对象的引用，可以直接使用。如果在 onCreate 方法之外，需要用到 SQLiteOpenHelper 类中提供的两个方法 getReadableDatabase 和 getWritableDatabase。这两个方法都用于打开或创建一个 SQLite 数据库(如果该数据库不存在则创建)，不同之处在于分别用于打开得到可读和可写的数据库。该对象除了提供 execSQL 直接执行 SQL 语句之外，还分别对增、删、改提供了相应的方法 insert、delete、update。插入数据代码如下：

```
public class MainActivity extends AppCompatActivity {
      private GameDBHelper gameDBHelper;
      private static final String TAG = "LogTest";
      @ Override
      protected void onCreate (Bundle savedInstanceState) {
           super. onCreate (savedInstanceState);
           View gameView = new View (this);
           setContentView (gameView);
           gameDBHelper = new
GameDBHelper (this,GameConst. DataStorage. DB_NAME);

           final TextView txtHello = (TextView) findViewById (R. id.
TxtViewHello);
           Button btnHello = (Button) findViewById (R. id. btnHello);

           btnHello. setOnClickListener (new View. OnClickListener () {
                @ Override
                public void onClick (View v) {
                      SQLiteDatabase db = gameDBHelper.
                        getWritableDatabase ();
                      ContentValues cv = new ContentValues ();
                      cv. put ("UserName","zhangsan");
                      cv. put ("UserLevel",1);
                      db. insert (GameConst. DataStorage. TABLE_NAME,null,cv);
                }
           });
      }
}
```

contentValues 用于保存键值对的数据。用于 SQLite 数据时，key 要与数据表中字段名相对应，value 则是要插入的值。Insert 方法的第一个参数为表名；第二个参数是当 contentValues 没有值时，由于 SQLite 不允许插入空行，因此将该参数指定的字段设为 null 值来完成数据插入；第三个参数就是 contentValues，代表要插入的数据。注意，在增删改操作时，都需要调用 getWritableDatabase 来获得可写入的数据库对象。后面查询代码则使用 getReadableDatabase 获得可读数据库对象。相应的删除代码方法为

```
db.delete(String table, String whereClause, String[] whereArgs);
```

该方法第一个参数同样为表名，第二个是 Where 条件，例如“UserName = ?”，其中“?”是占位符，其值由第三个参数指定，条件中可能有多个占位符，所以第三个参数为字符串数组，可以对应多个占位符。删除代码如下：

```
SQLiteDatabase db = gameDBHelper.getWritableDatabase();
String[] args = {String.valueOf("zhangsan")};
db.delete(GameConst.DataStorage.TABLE_NAME,"UserName = ?",args);
```

而修改的方法为

```
db.update(String table, Contentvalues values, String whereClause, String
whereArgs)
```

第一个参数为表名，第二个为修改的字段数据值，后两个参数与 delete 方法相同。实现代码如下：

```
SQLiteDatabase db = gameDBHelper.getWritableDatabase();
ContentValues cv = new ContentValues();
cv.put("UserName","zhangsan");
cv.put("UserLevel",2);
String[] args = {String.valueOf("zhangsan")};
db.update(GameConst.DataStorage.TABLE_NAME,cv,"UserName = ?",args);
```

8.3.3 SQLite 数据查询

查询方法有四种，定义如下：

```
db.rawQuery(String sql, String[] selectionArgs);

db.query(String table, String[] columns, String selection, String[]
selectionArgs, String groupBy, String having, String orderBy);

db.query(String table, String[] columns, String selection, String[]
selectionArgs, String groupBy, String having, String orderBy, String
limit);

db.query(String distinct, String table, String[] columns, String selection,
String[] selectionArgs, String groupBy, String having, String orderBy,
String limit);
```

第一个方法只有两个参数，分别是查询的 SQL 语句以及对应其中占位符的参数值，后面是三个重载方法，table 是表名，columns 是查询的字段，字段可以多个，是数组类型，selection 是查询条件，selectionArgs 是对应占位符的值，groupBy 是分组，orderBy 是排序，having 是分组的筛选，limit 是返回记录数限制，最后一个方法中的 distinct 表示是否有重复记录。

上述四个方法都是返回 Cursor，代表数据集的游标对象，取得多条记录时，通过移动游标指针到对应行，再根据字段读取数据。相应操作方法比较多，其中一部分常用方法如表 8 -4 所示。

表 8 -4 Cursor 操作方法

方法类别	方 法	说 明
游标位置操作	isAfterLast( )	游标指针是否在最后一条记录之后
	isBeforeFirst( )	游标指针是否在最前一条记录之前
	move(int offset)	移动指针，offset 是偏移值
	moveToFirst( )	指针移动到第一条记录
	moveToLast( )	移动到最后一条记录
	moveToNext( )	移动到下一条记录
	moveToPrevious( )	移动到前一条记录
	moveToPosition(int position)	移动到指定位置
读取数据操作	getColumnCount( )	获得列数(字段数)
	getColumnIndex(String columnName)	通过字段名获得字段索引
	getCount( )	获得记录条数
	getXXX(int columnIndex)	XXX 代表相应数据类型，例如 getString、getInt，该方法用于读取不同类型的字段值

查询的代码如下：

```
SQLiteDatabase db = gameDBHelper.getWritableDatabase();
ContentValues cv = new ContentValues();
cv.put("UserName", "zhangsan");
cv.put("UserLevel", 1);
db.insert(GameConst.DataStorage.TABLE_NAME, null, cv);
String[] columns = new String[]{"UserName","UserLevel"};
Cursor cursor = db.query(GameConst.DataStorage.TABLE_NAME, columns, "name
like ?", new String[]{"% zhang% "}, null, null, "UserLevel asc");
while (cursor.moveToNext()) {
    Log.i(TAG, cursor.getString(0));
    Log.i(TAG,cursor.getString(1));
}
cursor.close();
```

8.3.4 自定义 SQLite 访问类

为了在项目中访问 SQLite 数据库，本节将上面 SQLite 操作的内容进行封装。为了便于对 SQLite 的操作，首先需要对数据库中表的数据信息进行描述，这里通过定义相应的合同 Contract 类来完成，例如表 8 – 5 所示的数据表。

表 8 – 5 testTable

列 名	类 型	备 注
_ID	Integer	主键
columnName1	Integer	
columnName2	String	

对于一个数据表而言，需要记录其表名、列名及对应类型。针对上述 testTable，可以定义相应的合同类 testContract。该类中的成员都声明为静态，不需要在项目中对该类实例化，直接调用其静态方法即可。该类记录表的数据，如果需要在代码中对数据库中的表进行修改，直接修改该类即可。类的定义如下：

```
public class TestContract {
    public static final String TABLE_NAME = "testTable";

    public static final String SQL_CREATE_TABLE = "CREATE TABLE "
        + TABLE_NAME + "("
        + TestEntry._ID + " INTEGER PRIMARY KEY AUTOINCREMENT,"
        + TestEntry.COL_NAME1 + " INTEGER,"
        + TestEntry.COL_GTYPE + "TEXT);";

    public static abstract class TestEntry implements BaseColumns {
        public static final String COL_NAME1 = " columnName1";
        public static final String COL_NAME2 = " columnName2";
    }
}
```

该类中包含一个静态抽象内部类 TestEntry，用于记录列的名称。该类继承自 BaseColumns，BaseColumns 已包含_ID 列，所以如上述代码，数据表中共有三列，列名分别为_ID、columnName1、columnName2。然后在 Test 中添加成员保存表名以及创建表的 SQL 语句。有合同类读取它的信息进行数据库的操作，再定义一个 DatabaseHelper 继承 SQLiteOpenHelper 来完成这一步，代码如下：

```
public class DatabaseHelper extends SQLiteOpenHelper {
        public static final  int DB_VERSION = 1;

    public DatabaseHelper(Context context) {
        super(context, GameConst.DataStorage.DB_NAME,null,DB_VERSION
);
    }

    @ Override
    public void onCreate(SQLiteDatabase db) {
        db.execSQL(TestContract.SQL_CREATE_TABLE
);
    }

    @ Override
    public void onUpgrade ( SQLiteDatabase db, int oldVersion, int
    newVersion) {
        db.execSQL("DROP TABLE IF EXISTS" + TestContract.TABLE_NAME);

        onCreate(db);
    }

    public void initDBData(SQLiteDatabase db)
    {
        ContentValues testValue = new ContentValues();
        testValue.put(TestContract.TestEntry.COL_NAME1, 1);
        bulletValue.put(TestContract.TestEntry.COL_NAME2, "TEST STRING");
        db.insert(TestContract.TABLE_NAME, null, bulletValue);
    }
}
```

其中 onCreate 和 onUpgrade 是重写父类的方法，前者只有在数据库创建时会执行，换言之，当调用 getReadableDatabase( ) 或 getWritableDatabase( ) 发现数据库不存在时会执行，而如果此时数据库已经存在则不会执行。因此，测试时，只有第一次运行会执行；而数据库创建后，第二次运行项目，就不会运行。如果需要测试该方法，则需要在模拟器或真机中删除相应的数据库。文件位于“data \ data \ 包名 \ databases”文件夹下，模拟器可以通过打开 Android device monitor \ file explorer，找到该文件夹，将数据库删除即可，点击右上角的“ - ”按钮，如图 8 - 1 所示。

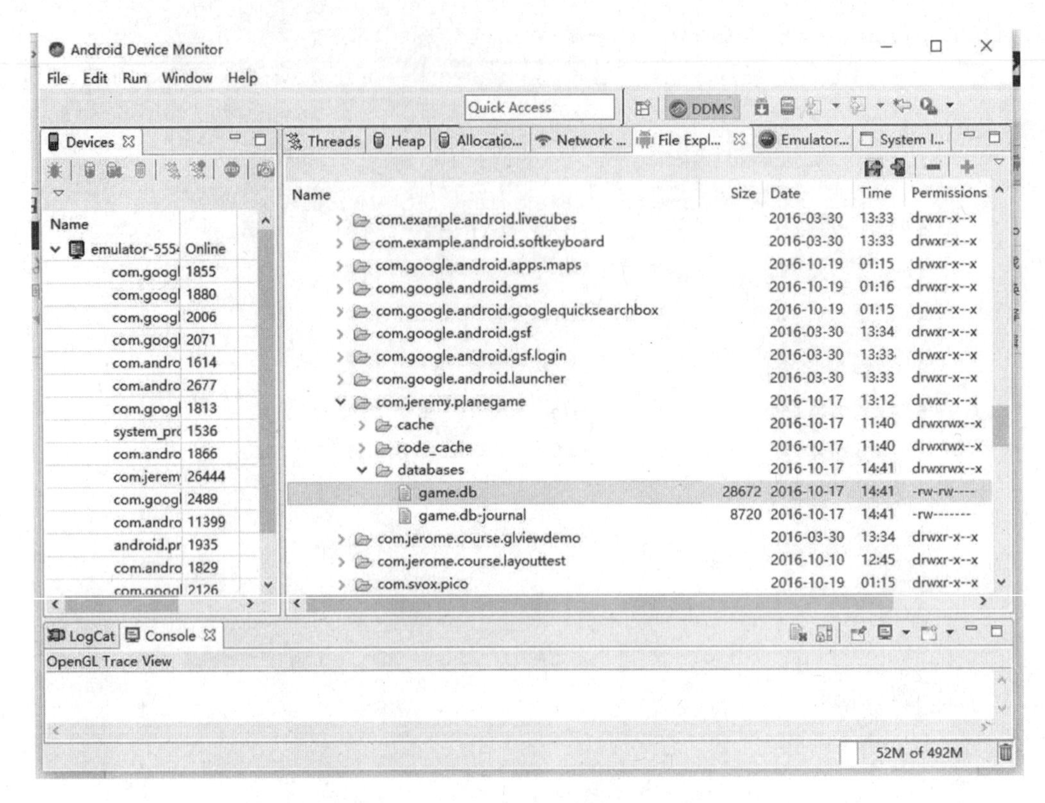

图 8－1 file explorer 模拟器的文件管理

onUpgrade 方法在数据库版本发生改变时会调用，这里版本号可以通过创建 DBHelper 时指定。在上述代码中，该方法先将之前的表删除，然后重新创建。

9 多线程

Android 启动时开启的线程称为 UI 主线程。在第 7 章中使用的 GLSurfaceView 会自带启动一个渲染线程。UI 主线程要做的工作很多，包括绝大部分事件的处理、UI 更新绘制等等。如果在 UI 线程插入太多游戏相关的其他工作，会使得应用无法及时地响应事件处理，给用户带来糟糕的体验。因此，为了让应用运行更加流畅，需要使用多线程，为操作复杂耗时的任务开启子线程，从而提升应用的并发执行性能。通常 Android 实现多线程处理方式有两种：AsyncTask 异步任务类和 Handler 机制。

9.1 AsyncTask

AsyncTask 是由 Android 框架提供的异步处理类，它将线程的开启和运行通信进行了封装，使用比较简单，适用于处理比较简单、生命周期较短的任务。AsyncTask 类定义如下：

```
public abstract class AsyncTask<Params, Progress, Result>
```

可以看到该类需要三个泛型类型，其中 Params 表示启动任务的输入参数，Progress 表示后台任务执行进度，Result 表示执行的结果类型。若没有则传入 java. lang. Void，注意不是空类型 void。启动一个异步任务，通常做法可以定义一个 AsyncTask 的子类，这个抽象类中可能需要重写的方法有：

- onPreExecute( )　当异步任务启动后，就会执行该方法。如果有需要，可以在这里显示任务开始提示之类，例如一个风火轮或进度条。
- doInBackground(Params··· params)　这是 AsyncTask 的抽象方法，也是唯一必须重写的方法，在 onPreExecute 执行后执行该方法。从命名上可以看出，该方法在后台执行，可以在此做一些比较耗时的操作。但它不是运行在 UI 主线程，所以不能访问 UI 信息。在执行过程中可以调用 publishProgress 来更新任务的进度。
- onProgressUpdate(Progress···)　如果调用了 publishProgress，就会执行此方法。它是在主线程执行，可以用于改变进度条的值，让用户掌握任务进度。
- onPostExecute(Result)　当 doInBackground 结束时，执行该方法。这里可以获得在 doInBackground 得到的结果，从而处理操作 UI。此方法在主线程执行，任务执行的结果作为此方法的参数返回。
- onCancelled( )　当用户调用取消时，想要执行的代码写在这里。

以上方法中 doInBackground 是抽象方法，必须实现，其他的可选。要注意的是，创建异步任务只能在 UI 主线程中。异步任务的启动通过 execute 方法来进行。该方法也只能在 UI 主线程调用。以下是使用 AsyncTask 的代码：

```
class DataReadAsyncTask extends AsyncTask <Context,Void,Void> {

    @ Override
    protected Void doInBackground(Context... params) {
        GameDBHelper gameDBHelper = new
GameDBHelper(params[0],GameConst.DataStorage.DB_NAME);
    SQLiteDatabase db = gameDBHelper.getWritableDatabase();
     ContentValues cv = new ContentValues();
     cv.put("UserName", "zhangsan");
     cv.put("UserLevel", 1);
      db.insert(GameConst.DataStorage.TABLE_NAME, null, cv);
      String[] columns = new String[]{"UserName","UserLevel"};
      Cursor cursor = db.query(GameConst.DataStorage.TABLE_NAME, columns,
"name like ?", new String[]{"% zhang% "}, null, null, "UserLevel asc");
      while (cursor.moveToNext()) {
            Log.i("", cursor.getString(0));
            Log.i("",cursor.getString(1));
      }
      cursor.close();
    return null;
   }
}
```

上述代码将插入 SQLite 数据的工作放到了异步任务之中，实际上创建 SQLite 也在这里，这样就要传入 context(设备上下文)。注意，AsyncTask 第一个泛型数据 params 即是可以传入的参数，通过 execute 执行时传入。该参数也自动作为 doInBackground 的参数。这里传入的是一组参数，所以用数组方式 params[0]访问唯一的参数。创建该类后，可以在 Activity 中构建实例并运行。代码如下：

```
DataReadAsyncTask at = new DataReadAsyncTask();
at.execute(this);
```

如果异步任务比较简单，也可以直接作为匿名内部类的方式在 Activity 类中定义并实现。AsyncTask 适合做一些简单的一次性的异步任务，例如游戏开始前读取数据需要显示进度条的任务，但较复杂并持续时间长的任务就不适合。

9.2 Handler机制

在应用开发中采用多线程，线程需要接受任务执行并与其他的线程进行通信、协作。其中一些线程可能还需要与UI打交道，通知UI主线程更新。Android系统提供Handler机制处理线程之间的协作通信过程。

9.2.1 Handler消息处理

完整的Handler机制涉及Handler、Message、MessageQueue、Looper和HandlerThread。其中Handler负责发送和处理消息，从而实现线程间的通信。Looper则负责管理线程的消息队列和消息循环。Message是消息的载体，存放消息；MessageQueue是消息队列，存放等待处理的消息。而HandlerThread以及前一节的AsyncTask都是对Handler的封装使用。

Handler发送消息的方法部分列举如下：

- sendEmptyMessage(int)；　发送空消息
- sendMessage(Message)；　发送消息
- sendMessageAtTime(Message, long)；　指定时间点发送消息
- sendMessageDelayed(Message, long)；　延迟多长时间后发送消息

除了发送消息，Handler还可以启动某工作任务(runnable)，实际上也是通过发送消息完成的。方法如下：

- post(Ruannable)；　立即启动工作任务
- postAtTime(Runnable, long)；　指定时间点启动该工作任务
- postDelayed(Runnable long)；　延长多长时间启动该工作任务

举例说明Handler的工作机制：有一个繁琐的工作任务(读取数据库的数据)，那么在UI主线程中创建一个子线程，由它完成读取工作，完成之后需要在UI界面做出提示，用户获知任务完成。而子线程不能更新UI，因此在主线程中创建Handler实例，并重写其处理XX消息的方法，在其中更新UI。因为Handler是主线程的，所以可以更新UI。在子线程任务完成后通过Handler插入一条XX消息到主线程的消息队列。消息队列中的消息会通过Looper依次处理。当处理到XX消息时，UI更新，代码如下：

```
final Handler handler = new Handler(){
      public void handleMessage(Message msg)
      {
            switch (msg.what)
            {
                  case 0:
                        Log.i(TAG,"thread run!");
            }
```

```
        }
};

new Thread(new Runnable() {
        @ Override
        public void run() {
                handler.sendEmptyMessage(0);
        }
}).start();
```

上述代码位于 Activity 中，在 UI 主线程创建了一个 Handler 实例，并重写了 HandleMessage 处理消息这个方法。在随后代码中启动了一个子线程，子线程中调用 sendEmptyMessage 发送一个消息，其中参数就是发送的 message 的 what 成员。在开发中，往往通过 what 值来区分不同的消息类型。

以上并没有出现 Looper 和 MessageQueue，当然消息队列 MessageQueue 是由 Looper 维护的，无需程序员手动操作，但 Handler 必须有 Looper 才能工作。实际上作为 UI 主线程，系统已经为它创建 Looper，所以在 UI 主线程中可以直接创建 Handler 实例。构建 Handler 实例时，会自动获取到当前线程的 looper。如果在一个没有 Looper 的线程构建 Handler，则会抛出异常。例如，下面代码运行会产生异常：

```
new Thread(new Runnable() {
    @ Override
    public void run() {
        Handler handler = new Handler(){
            public void handleMessage(Message msg)
            {
                Log.i(TAG,"main thread send msg");
            }
        };
    }
}).start();
```

新建的子线程需要创建 Looper 才能在其中创建 Handler 实例。Looper 创建启动只需要用到两个简单方法——prepare 和 loop，前者是创建 Looper，后者是建立消息循环，就是让 Looper 工作起来。一个线程中只能有一个 Looper、一个消息队列，所以 prepare 在一个线程中只能执行一次。将上述产生异常代码改为如下代码即可正常运行。

```
new Thread(new Runnable() {
    @ Override
    public void run() {
        Looper.prepare();
         MainActivity.this.handler = new Handler(){
```

```
                public void handleMessage(Message msg)
                {
                        Log.i(TAG,"main thread send msg");
                }
            };
            Looper.loop();
    }
}).start();
```

一个线程中只能有一个Looper，但可以创建多个Handler，每个Handler构建时默认关联自身所在线程的Looper，但也可以指定其他线程的Looper。Handler发送消息后，该消息处理运行在Handler关联的Looper所在的线程。

9.2.2 HandlerThread

如果觉得上一节创建线程后还需要创建Looper才能启用Handler的过程繁琐，Android提供了封装好的线程HandlerThread。从名字上可以看出，它是个线程，有别于一般线程，它已经有了Looper。如果有一些工作要完成，直接创建HandlerThread和关联其Looper的Handler即可。代码如下：

```
workThread = new HandlerThread("WorkerThread");
workThread.setPriority(Thread.MIN_PRIORITY);
workThread.start();
workHandler = new Handler(workThread.getLooper());
Runnable workTask = new Runnable() {
    @Override
    public void run() {
        //执行工作任务
    }
};
workHandler.post(workTask);
```

如果应用的一些工作需要在后台运行，完成后更新UI，使用HandlerThread实现，那么首先需要创建HandlerThread的实例，其次需要两个Handler，一个关联UI主线程的Looper(在主线程中使用无参构造函数构建实例即可)，该Handler用于工作完成后更新UI；另外一个则关联HandlerThread的Looper，该Handler便于主线程传递工作。代码如下：

```
private Handler uiHandler;
private Handler workHandler;
HandlerThread workThread;
@Override
protected void onCreate(Bundle savedInstanceState) {
```

```
        super.onCreate(savedInstanceState);
        View gameView = new GameGLView(this);
        setContentView(gameView);

        uiHandler = new Handler();
        workThread = new HandlerThread("working");
        workThread.start();
        workHandler = new Handler(workThread.getLooper())
        {
            @ Override
             public void handleMessage(Message msg)
            {
                //耗时工作代码
                //...
                uiHandler.post(new Runnable() {
                    @ Override
                    public void run() {
                         //更新 ui 代码
                         //...
                    }
                });
            }
        };
    }
```

与 AsyncTask 相比，HandlerThread 可以持续工作。AsyncTask 接受任务，迅速开启工作线程，完成返回结果；而 HandlerThread 则是长时间持续运行，可以不断接受任务进行处理，处理完等待下一个任务。在实际开发中，根据任务的需求目的做不同的选择。

9.3 ThreadPool

对于 Android 客户端编程，尤其在游戏中，应用线程池的机会并不多。例如网络请求，设立一个网络线程来与服务器通信即可，但并不能满足所有情况，如从网络上请求很多张图片资源时，多个线程会比单个线程更高效。线程数量并非越多越好，每个线程创建销毁也都占用时间，线程一多，线程间调度的时间可能会造成性能下降。当需要线程比较多时，采用线程池技术是必要的，JDK 提供了线程池的实现类 ThreadPoolExecutor。创建线程池通常调用 Executors 的工厂方法来创建 ThreadPoolExecutor 实例。下面三个方法用于创建不同类型的线程池。

- Executors.newFixedThreadPool　该方法创建一个可重用固定线程数的线程池，以共享的无界队列方式来运行这些线程。
- Executors.newSingleThreadExecutor　创建一个使用单个 worker 线程的 Executor，以

无界队列方式来运行该线程。

- Executors. newCachedThreadPool　创建一个可根据需要创建新线程的线程池，但是在以前构造的线程可用时将重用它们。对于执行很多短期异步任务的程序而言，这些线程池通常可提高程序性能。

线程池创建之后，通过调用 execute 来执行任务，这个过程比想象中简单得多。代码如下：

```
threadPool = Executors.newFixedThreadPool(20);
Runnable workTask = new Runnable() {
    @Override
    public void run() {
        //执行工作任务
        Log.i(TAG,Thread.currentThread().getName());
    }
};
for(int i=0;i<20;i++)
{
    threadPool.execute(workTask);
}
```

ThreadPoolExecutor 提供了 shutdown 方法来关闭线程池，该方法会等待线程池完成所有现有任务之后关闭。

9.4　线程优先级

如果线程同时存在数量过多，超过 CPU 能同时执行的最大线程数，CPU 会根据线程的优先级进行调度切换，不同的优先级获得不同的时间片。Android 系统中，前台进程会比后台进程获得更多的时间片。而当创建线程时，线程默认的优先级是和所在母线程保持一致，所以在 UI 主线程中，创建线程时应该为它设定更低的优先级，避免其与主线程抢夺资源。

优先级的设置有两种方式，一种是通过 Thread 的方法 setPriority，有定义的三个常量分别为：

```
public static final int MAX_PRIORITY = 10;
public static final int MIN_PRIORITY = 1;
public static final int NORM_PRIORITY = 5;
```

除了三个常量，可以直接用 int(1～10)表示，1 优先级最低，10 优先级最高。

还有一种方式是通过 android. os. Process 的 setThreadPriority，有两个同名的重载方法，一个用于指定线程的优先级，一个用于指定当前线程的优先级，取值范围在(-20～19)。和前一种方法相反，该值越小，表示优先级越高，例如设为 19 的线程优先级最低。

10 网络通信

随着移动互联网的高速增长，移动平台上游戏 App 也都从单机向网络化方向发展，即使是单机游戏也需要实现各类的网络服务，如游戏高分排行榜等。游戏中的网络通信通常都通过 Socket 进行开发。

10.1 Socket 通信

Android 涉及网络服务的 App 与服务器通信的方式不外乎两种，即 Http 和 Socket。其中 Http 连接采用“request – response”，每次客户端向服务器发送请求，服务器响应。这种方式明显不适用于网络游戏。服务器需要主动向客户端发送消息，而 Socket 则在双方建立连接，进行数据传输，客户端与服务端均可以向对方主动发送数据。游戏应用中网络通信主要按照 C/S 架构来搭建，包括服务端和客户端两个部分。

服务端工作步骤如下：启动服务器，监听指定端口，等待客户端连接请求，接收到连接请求，建立连接，传输数据，关闭连接。

客户端的工作步骤如下：打开客户端，发送连接请求，建立连接，传输数据，关闭连接。两者通信如图 10 – 1 所示。

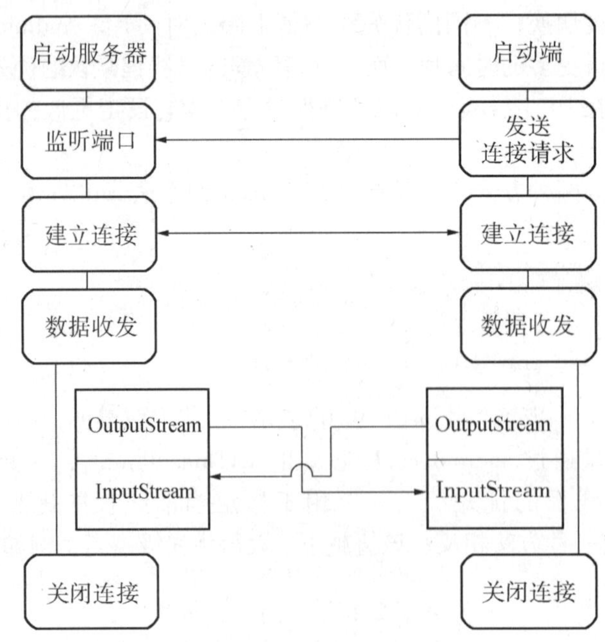

图 10 – 1　网络数据传输

10.1.1 服务端编程

服务器端的开发可以根据服务器情况选择相应的语言，在这里，考虑为便于学习，采用 Java 语言编写。在服务器开启一个可变数量线程池，为每一个客户端连接分配一个线程来处理，收到客户端请求，在控制台输出客户端 IP 地址信息，代码如下：

```
public static void main(String[] args) {
        //声明一个 ServerSocket 对象
        ServerSocket serverSocket = null;
        try {
            serverSocket = new ServerSocket(2015);
            System.out.println("等待客户端连接...");
            ExecutorService threadPool = Executors.newCachedThreadPool();
            boolean bRun = true;
            while(bRun) {
                Socket socket = serverSocket.accept();
                System.out.println(socket.getInetAddress().getHostAddress
() + "客户端连接");
                threadPool.execute(new DoClientTask(socket));
            }
            serverSocket.close();
        } catch (IOException e) {
            e.printStackTrace();
        }
}
```

为每个客户端线程启动一个处理线程，线程工作代码如下：

```
public class DoClientTask implements Runnable {
    private Socket socket;
    private static final int BUFFERSIZE = 1024;

    public DoClientTask(Socket socket)
    {
        this.socket = socket;
    }

    @Override
    public void run()
    {
        try
```

```
            {
            DataInputStream inStream = new
DataInputStream(socket.getInputStream());
            DataOutputStream outStream =
new DataOutputStream(socket.getOutputStream());
            outStream.writeUTF("welcome to 2015");
            int recvSize = 0;
            byte[] buf = new byte[BUFFERSIZE];
            while((recvSize = inStream.read(buf))! = -1)
            {
                  outStream.write(buf,0,recvSize);
            }
        } catch (IOException e) {
            e.printStackTrace();
        }finally {
            try {
                  socket.close();
            } catch (IOException e) {
                  e.printStackTrace();
            }
        }
    }
}
```

上面服务器处理客户端连接启动一个循环，从 socket 读数据，每次循环简单地把客户端发送过来的数据发回去，具体的处理要根据游戏应用具体需求。

10.1.2 客户端编程

Android 客户端要进行网络连接，也需要申请网络权限，在 manifest 中添加访问权限代码：

```
<uses-permission android:name="android.permission.INTERNET" />
```

目前功能实现比较简单，收到服务器发送过来的数据，就在 Log 中输出收到数据长度。代码如下：

```
public class ReceiveData implements Runnable {
    private Socket socket;
    private static final int BUFFERSIZE = 1024;
    private Handler netHandler;

    private DataInputStream inStream ;
    private DataOutputStream outStream ;
```

```
    private boolean bRun;
    public ReceiveData(Socket socket,Handler netHandler)
    {
        this.socket = socket;
        this.netHandler = netHandler;
    }
    @Override
    public void run() {
        try {

            inStream = new DataInputStream(socket.getInputStream());
            outStream = new DataOutputStream(socket.getOutputStream());
            bRun = true;
            int recvSize = 0;
            byte[] buf = new byte[BUFFERSIZE];

            while((recvSize = inStream.read(buf))! = -1)
            {
                Log.i("NET","收到数据"+recvSize);
                netHandler.sendEmptyMessage(0);
            }
        } catch (IOException e) {
            e.printStackTrace();
        }
    }
}
```

上述代码定义了一个类实现 Runnable 接口。该类 run 方法中连接服务器并接受服务器发送的数据，然后输出收到数据长度，为网络操作定义一个 NetProxy 类，完成连接服务器，成功后创建一个 HandlerThread 实例来接收服务器数据，然后通过 Handler 在该线程中启动连接服务器的任务，并在收到服务器数据后通过 Handler 通知主线程。代码如下：

```
public class NetProxy {
    private HandlerThread netThread;
    private Handler receHandler;
    private Handler netHandler;
    private String ip;
    private int port;

    private Socket socket;
```

```
    private boolean bConnect = false;

    public NetProxy() {
    }

    public boolean isConnected() {
        return bConnect;
    }
    public void setConnected(boolean connected) {
        this.bConnect = connected;
    }

    public void connectServer()
    {
        try {
            socket = new Socket("localhost",2015);
        } catch (IOException e) {
            e.printStackTrace();
            this.setConnected(false);
        }
        this.setConnected(true);

        netThread = new HandlerThread("net");
        netThread.start();
        receHandler = new Handler(netThread.getLooper());
        netHandler = new Handler(){
            @Override
            public void handleMessage(Message msg)
            {
                Log.i("NET", "收到数据" + msg.what);
            }
        };
        ReceiveData receiveData = new ReceiveData(socket,netHandler);
        receHandler.post(receiveData);
    }

    public boolean disConnect()
    {
        if(socket == null)
        {
            return true;
        }
```

```
            try {
                socket.close();
            } catch (IOException e) {
                e.printStackTrace();
            }
            this.setConnected(false);
            return true;
        }

        public void sendData()
        {
        }
    }
```

10.2 游戏网络数据处理

10.1 节代码实现了简单的服务器端和客户端，功能有限，服务器将每个客户端发送过来的数据原封不动发回去，而客户端则简单输出服务器发过来的字节长度。但实际上两者之间的数据往来很频繁且种类众多，对应不同的数据应做不同的处理。

10.2.1 Serializable 序列化

如果服务器也是用 Java 语言编写，那么通过 Serializable 来传输对象数据是不错的选择。Serializable 是 Java 提供的序列化接口，Java 可以将实现该接口的对象转换成字节序列，并且这个过程可逆，所以可以将对象数据进行网络传输、文件保存等。

在网络传输中，只需将 socket 的 inputStream 和 outputStream 分别用 ObjectInputStream 和 ObjectOutputStream 包装，即可分别调用 readObject 和 writeObject 进行对象数据的读写。对象序列化网络传输如图 10－2 所示。

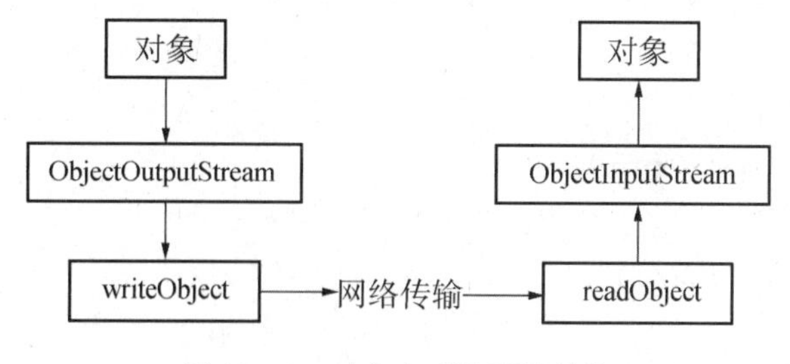

图 10－2　对象序列化网络传输

下面代码从客户端发送一个对象数据给服务器端，当然两边都需要同样的类定义。类定义代码如下：

```
public class NetData implements Serializable{
     private int userID;
     private int userLevel;
     public void NetData(int id,int level)
     {
        userID = id;
        userLevel = level;
     }
        public String toString()
{
     return "userID:" + userID  + " userLevel:" + userLevel;
}

}
```

客户端代码：

```
NetData netData  = new NetData(1,10);
ObjectOutputStream oos  = new ObjectOutputStream(socket.getOutputStream());
oos.writeObject(netData);
```

服务器端代码：

```
ObjectInputStream ois  = new ObjectInputStream(socket.getInputStream());
NetData netData  =  (NetData)ois.readObject();
System.out.println(netData.toString());
```

10.2.2　JSON格式

JSON全称JavaScript Object Notation，是一种轻量级的数据交换格式，采用与编程语言无关的文本格式，在服务器端和客户端采用不同语言的情况下也不影响数据的交换。在前面的Android开发学习中，可以看到Android项目中大量应用了XML文件来描述数据，如manifest和布局文件等。而JSON与XML相比，更加简单、数据体积小、易于传输，解析也更加方便。下面是一个JSON数据例子。

```
var userInfo =
          {
          "nickname": "bigHero",
          "level" : 10,
          "id": 1,
          "regInfo" : {
          "name" : "zhangsan",
          "phone" : "88888888",
          "city" :  "guangzhou"
          }
 }
```

无论是XML，还是JSON，都是文本方式存储，传输时可以将其转换为字符串，而接收时可以将字符串转回JSON对象，这是一个编码和解码的过程。

10.2.3 Byte传输

JSON是通过字符形式传输的。如果对效率有更高要求，应该采用bytes字节传输方式，可以通过对服务器和客户端定义同样的数据格式，保持收发的数据格式一致。这样，在发送时将数据转换成字节流，在读取时按照同样的顺序再进行解析即可。在socket连接中，可以采用UDP和TCP两种方式，TCP采用安全连接，不存在丢包的风险，但实际运行中，网络环境比较复杂，会有粘包可能。简单地说，发送的两次数据可能会一次收到，这时需要对包进行解析。简单的处理办法是，在发送数据前，先发送本次要发送数据的长度，这样，在读取时先获取长度，然后再根据长度检查缓冲区中数据是否足够，如果足够，则读取相应长度的数据，多余的作为下一条数据处理。

由于网络的数据时刻处在监听状态，因此socket的缓冲区数据要取出来。对于网络线程而言，其并不负责数据的解析工作，只是将数据取出来，然后专心下一条数据的接收，所以需要创建新的缓冲区，用来接收网络线程取出来的数据，然后由其他线程来解析。这里我们定义一个环形缓冲来保存数据，避免假溢出的情况。实际上可通过数据结构的循环队列来实现，代码如下：

```
public class RingBuffer {
      private byte[] buf = null;
      private int writeIndex;
      private int readIndex;

      public RingBuffer(int len)
      {
        buf = new byte[len];
        writeIndex = readIndex = 0;
      }

      public synchronized boolean put(byte[] bytes)
      {
        //留一个数组空位不用,这样队列空和满的判断条件就不一样
        int emptylen = (buf.length + (readIndex - writeIndex) - 1)% buf.length;
        if(emptylen == 0)// (writeIndex + 1)% buf.length == readIndex)//队列满
        {
            return false;
        } else if (bytes.length > emptylen) {
            return false;
        }else if (bytes.length > buf.length - writeIndex)
         //只判断 write 距离最后距离是否够,不用考虑 read
        {
```

```
            System.arraycopy(bytes, 0, buf, writeIndex, buf.length -
writeIndex);
            System.arraycopy(bytes, buf.length - writeIndex, buf, 0,
bytes.length - buf.length + writeIndex);
            writeIndex = bytes.length - buf.length + writeIndex;
        } else {
            System.arraycopy(bytes, 0, buf, writeIndex, bytes.length);
            writeIndex = (bytes.length + writeIndex) % buf.length;
        }
        return true;
    }

    public synchronized boolean put(byte[] bytes,int len)
    {

        int emptylen = (buf.length + (readIndex - writeIndex) - 1)% buf.length;
        //留一个数组空位不用
        if (emptylen ==0)// (writeIndex +1)% buf.length == readIndex)//队列满
        {
            return false;
        } else if (len > emptylen) {
            return false;
        } else if (len > buf.length - writeIndex)
        //只判断 write 距离最后距离是否够,不用考虑 read
        {
            System.arraycopy(bytes,0, buf,writeIndex,buf.length - writeIndex);
            System.arraycopy (bytes, buf.length - writeIndex, buf, 0, len -
buf.length + writeIndex);
            writeIndex = len - buf.length + writeIndex;
        } else {
            System.arraycopy(bytes, 0, buf, writeIndex, len);
            writeIndex = (len + writeIndex) % buf.length;
        }
        return true;
    }

    public synchronized byte[] get(int len,boolean bMoveIndex)
    {
        int readablelen = ((writeIndex - readIndex) +buf.length)% buf.length;
        if(len > readablelen)
```

```
        {
            return null; //数据长度不足!
        }
        byte[] bytes = new byte[len];
        if(len <= buf.length - readIndex )
        //不用考虑write,如果write>read,那肯定也可以满足该条件
        {
            System.arraycopy(buf,readIndex,bytes,0,len);
            if(bMoveIndex) {
                readIndex = (readIndex + len) % buf.length;
            }
        }else
        {
            System.arraycopy(buf,readIndex,bytes,0,buf.length -
readIndex);
            System.arraycopy(buf,0,bytes,buf.length - readIndex,len -
buf.length + readIndex);
            if(bMoveIndex) {
                readIndex = (readIndex + len) % buf.length;
            }
        }
            return bytes;
        }

        public boolean hasReadableData(int len)
        {
            int readablelen = ((writeIndex - readIndex) + buf.length)%
buf.length;
            if(len > readablelen)
            {
                return false; //数据长度不足!
            }
            return true;
        }

        public byte[] get(int len)
        {
            return get(len,true);
        }

        public byte[] getData()
        {
```

```
            byte[] bytes = get(4,false);
            if(bytes == null)
                return null;
            int datalen = ByteUtil.BytesToInt(bytes);

            bytes = null;
            if(hasReadableData(datalen+4))
            {
                get(4);
                bytes = get(datalen);
            }

            return bytes;
        }

        public synchronized void clear()
        {
            Arrays.fill(buf, (byte)0);
            readIndex = writeIndex = 0;
        }

}
```

对于网络线程而言，不用进行解析工作，工作量比较简单，只负责接收数据，然后调用 RingBuffer 的 put 方法将取得的数据放入环形缓冲即可。

```
public class NetProxy extends Thread{//implements Runnable{

    private Handler mainHandler;
    private String ip;
    private int port;
    private Socket socket;
    private boolean bConnect = false;
    private DataInputStream inStream ;
    private DataOutputStream outStream ;
    private RingBuffer ringBuffer;

    public NetProxy(Handler mainHandler,RingBuffer buf) {
        this.mainHandler = mainHandler;
        ringBuffer = buf;
    }
```

```
    public boolean isConnected() {
        return bConnect;
    }
    public void setConnected(boolean connected) {
        this.bConnect = connected;
    }

    @Override
    public void run() {

        connectServer();

mainHandler.sendEmptyMessage(GameConst.HandleMsg.CONN_SUCCESS);

        int recvSize = 0;
            byte[] buf = new byte[1024];

        try {
            while((recvSize = inStream.read(buf))!=-1)
            {
                ringBuffer.put(buf,recvSize);
                //Log.i("NET","收到数据"+recvSize);

mainHandler.sendEmptyMessage(GameConst.HandleMsg.RECV_DATA);
        }
        } catch (IOException e) {
            e.printStackTrace();
        }

    }

    public void connectServer()
    {
        try {
            socket = new Socket("localhost",2015);
            inStream = new DataInputStream(socket.getInputStream());
            outStream =new DataOutputStream(socket.getOutputStream());

        } catch (IOException e) {
            e.printStackTrace();
            this.setConnected(false);
        }
```

```
            this.setConnected(true);
        }


        public boolean disConnect()
        {
            if(socket == null)
            {
                return true;
            }
            try {
                socket.close();
            } catch (IOException e) {
                e.printStackTrace();
            }
            this.setConnected(false);
            return true;
        }

        public void sendData(byte[] bytes)
        {
            if(!this.isConnected())
            {
                return;
            }
            try {
                outStream.write(bytes);
            } catch (IOException e) {
                e.printStackTrace();
            }
        }

}
```

解析数据时，从环形缓冲中先读取下一条数据的长度，然后再根据长度检测数据长度是否满足要求，满足则取得数据。

```
public class DataReader implements Runnable{

    private RingBuffer buf;
    private Handler mainHandler;

    public DataReader(Handler main,RingBuffer buf)
    {
        this.buf = buf;
        this.mainHandler = main;
    }

    @Override
    public void run() {
        byte[] bytes;

        while((bytes = buf.getData())!=null)
        {
            byte[] dataTypeBytes = new byte[4];
            System.arraycopy(bytes, 0, dataTypeBytes, 0, 4);
            int dataType = ByteUtil.BytesToInt(dataTypeBytes);

            byte[] dataBytes = new byte[bytes.length-4];
            System.arraycopy(bytes,4,dataBytes,0,bytes.length-4);

            Message msg = new Message();
            msg.what = dataType;
            msg.obj = bytes;

            mainHandler.sendMessage(msg);

        }
    }
}
```

11 游戏中的声音

一款可以发行的游戏，不能缺少声音。游戏中的声音可以分为两种，一是游戏中的背景音乐，二是游戏的即时音效。

11.1 MediaPlayer 音乐播放

MediaPlayer 是 Android 提供用于媒体播放的类，实际上不仅可以播放声音，还也可以播放视频，实现了一个多媒体播放器所用到的所有功能。游戏中的背景音乐比较长，通常用 MediaPlayer 来播放。

下面先介绍 MediaPlayer 的生命周期。MediaPlayer 的生命周期包括 10 个状态，每个状态都有对应可以进行的操作，各个状态之间转换如图 11 - 1 所示。

(1) Idle 状态：当使用 new( ) 方法创建一个 MediaPlayer 对象或者调用其 reset( ) 方法时，该 MediaPlayer 对象处于 idle 状态。这两种方法的一个重要差别是，如果在这个状态下调用了 getDuration( ) 等方法（相当于调用时机不正确），通过 reset( ) 方法进入 idle 状态时会触发 OnErrorListener. onError( )，并且 MediaPlayer 会进入 Error 状态；如果是新创建的 MediaPlayer 对象，则不会触发 onError( )，也不会进入 Error 状态。

(2) End 状态：通过 release( ) 方法可以进入 End 状态，只要 MediaPlayer 对象不再被使用，就应当尽快将其通过 release( ) 方法释放掉，以释放相关的软硬件组件资源，这其中有些资源是只有一份的（相当于临界资源）；如果 MediaPlayer 对象进入了 End 状态，则不会再进入任何其他状态。

(3) Initialized 状态：这个状态比较简单，MediaPlayer 调用 setDataSource( ) 方法进入 Initialized 状态，表示此时要播放的文件已经设置好。

(4) Prepared 状态：初始化完成之后还需要通过调用 prepare( ) 或 prepareAsync( ) 方法。这两个方法一个是同步的一个是异步的，只有进入 Prepared 状态，才表明 MediaPlayer 到目前为止都没有错误，可以进行文件播放。

(5) Preparing 状态：这个状态比较好理解，主要是与 prepareAsync( ) 配合，如果异步准备完成，会触发 OnPreparedListener. onPrepared( )，进而进入 Prepared 状态。

(6) Started 状态：显然，MediaPlayer 一旦准备好，就可以调用 start( ) 方法，这样 MediaPlayer 就处于 Started 状态，这表明 MediaPlayer 正在播放文件过程中。可以使用 isPlaying( ) 测试 MediaPlayer 是否处于 Started 状态。如果播放完毕，而又设置了循环播放，则 MediaPlayer 仍然会处于 Started 状态。类似地，如果在该状态下 MediaPlayer 调用了 seekTo( ) 或者 start( ) 方法，均可以让 MediaPlayer 停留在 Started 状态。

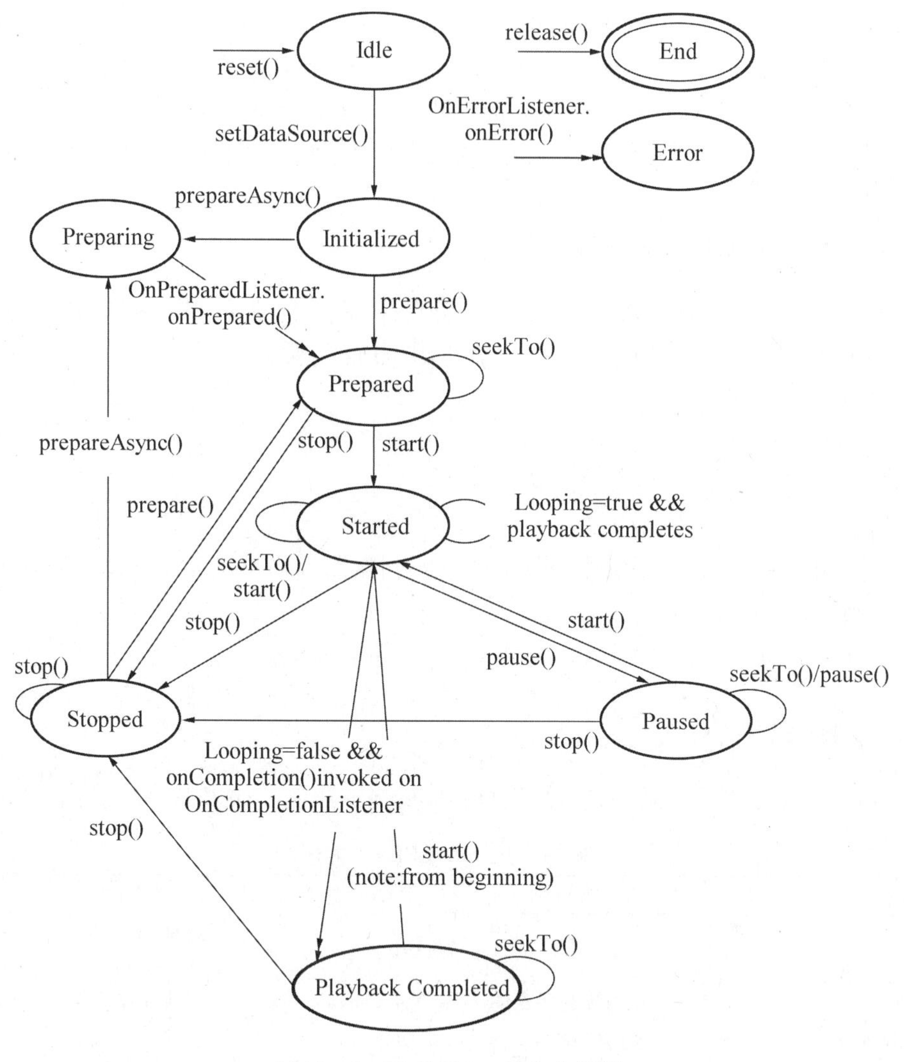

图 11－1　MediaPlayer 的生命周期

(7) Paused 状态：Started 状态下 MediaPlayer 调用 pause( ) 方法可以暂停 MediaPlayer，从而进入 Paused 状态。MediaPlayer 暂停后再次调用 start( ) 则可以继续 MediaPlayer 的播放，转到 Started 状态。暂停状态时可以调用 seekTo( ) 方法，这是不会改变状态的。

(8) Stop 状态：Started 或者 Paused 状态下均可调用 stop( ) 停止 MediaPlayer，而处于 Stop 状态的 MediaPlayer 要想重新播放，需要通过 prepareAsync( ) 和 prepare( ) 回到先前的 Prepared 状态重新开始才可以。

(9) PlaybackCompleted 状态：文件正常播放完毕而又没有设置循环播放，就进入该状态，并会触发 OnCompletionListener 的 onCompletion( ) 方法。此时，可以调用 start( ) 方法重新从头播放文件，也可以 stop( ) 停止 MediaPlayer，或者也可以 seekTo( ) 来重新定位播放位置。

(10) Error 状态：如果由于某种原因 MediaPlayer 出现了错误，会触发

OnErrorListener. onError( )事件，此时 MediaPlayer 即进入 Error 状态。及时捕捉并妥善处理这些错误很重要，可以帮助我们及时释放相关的软硬件资源，也可以改善用户体验。通过 setOnErrorListener(android. media. MediaPlayer. OnErrorListener) 可以设置该监听器。如果 MediaPlayer 进入了 Error 状态，可以通过调用 reset( )来恢复，使得 MediaPlayer 重新返回到 Idle 状态。

11.1.1 创建 MediaPlayer 实例

MediaPlayer 的创建方式有两种，一是直接调用构造函数，在构建实例后用 setDataSource 指定播放的数据来源。下面代码中的 path 变量存放媒体文件的文件路径，可以是本地路径，也可以是网络路径。

```
MediaPlayer mp = new MediaPlayer();
mp =MediaPlayer. setDataSource(path);
mp. prepare();
```

或者调用 MediaPlayer 提供的 create 方法进行构建，create 的第二个参数对应应用中自带的 resource 标识：

```
MediaPlayer mp = MediaPlayer. create(this, R. raw. bgmusic);
```

11.1.2 播放控制方法

MediaPlayer 提供了一系列方法来进行媒体的播放控制。如表 11 - 1 所示。

表 11 - 1 MediaPlayer 方法

方 法	说 明	方 法	说 明
MediaPlayer	构造方法	seekTo	指定播放的位置(以毫秒为单位的时间)
create	创建一个要播放的多媒体	setAudioStreamType	设置流媒体的类型
getCurrentPosition	得到当前播放位置	setDataSource	设置多媒体数据来源
getDuration	得到文件的时间	setDisplay	设置用 SurfaceHolder 来显示多媒体
getVideoHeight	得到视频的高度	setLooping	设置是否循环播放
getVideoWidth	得到视频的宽度	setOnButtering UpdateListener	网络流媒体的缓冲监听
isLooping	是否循环播放	setOnErrorListener	设置错误信息监听
isPlaying	是否正在播放	setOnVideoSize ChangedListener	视频尺寸监听
pause	暂停	setScreenOnWhile Playing	设置是否使用 SurfaceHolder 来保持屏幕显示
prepare	准备(同步)	setVolume	设置音量
prepareAsync	准备(异步)	start	开始播放
release	释放 MediaPlayer 对象	stop	停止播放
reset	重置 MediaPlayer 对象		

注意：如果 MediaPlayer 实例是由 create 方法创建的，那么第一次启动播放前不需要再调用 prepare( )，因为在 create 方法中已经调用过。

11.1.3 MediaPlayer 监听

MediaPlayer 提供了一些设置不同监听器的方法以更好地对播放器的工作状态进行监听，以期及时处理各种情况。常用的监听器注册方法如表 11 -2 所示。

表 11 -2 常用的监听器注册方法

方 法	说 明
setOnPreparedListener	视频准备完成后调用
setOnCompletionListener	视频播放完毕时调用
setOnErrorListener	异步操作过程出错时调用
setOnBufferingUpdateListener	网络缓冲变化时调用
setOnSeekCompleteListener	Seek 定位完成后调用
setOnVideoSizeChangedListener	视频尺寸改变时调用
setOnInfoListener	有提示信息时调用

设置播放器时需要考虑到播放器可能出现的情况，设置好监听和处理逻辑，以保持播放器的健壮性。

11.1.4 MediaPlayer 播放音乐

定义一个类用于控制游戏中声音播放，代码如下：

```
public class GameAudioManager {
    float volumeRatio;
    private List <MediaPlayer > mpList;

    public GameAudioManager()
    {
        resManager = ResManager.getInstance();
        mpList = new ArrayList <MediaPlayer > ();
    }

    public void playMusic(String musicResName)
    {
        MediaPlayer mp =
(MediaPlayer)resManager.getResForName(musicResName);
        mp.setOnErrorListener(new MediaPlayer.OnErrorListener() {
            @Override
```

```
                public boolean onError(MediaPlayer mp, int what, int extra) {
                    mp.reset();
                    return false;
                }
            });
            mp.start();
            mpList.add(mp);
        }

        public void pauseMusic(String musicResName)
        {
            MediaPlayer mp =
(MediaPlayer)resManager.getResForName(musicResName);
            mp.pause();
        }

        public void stopMusic(String musicResName)
        {
            MediaPlayer mp =
(MediaPlayer)resManager.getResForName(musicResName);
            mp.stop();
        }

        public void stopAllMusic()
        {
            Iterator<MediaPlayer> iter = mpList.iterator();
            while(iter.hasNext())
            {
                iter.next().stop();
            }
        }

        public void resumeMusic(String musicResName)
        {
            MediaPlayer mp =
(MediaPlayer)resManager.getResForName(musicResName);
            mp.start();
        }
}
```

在该类中，建立了一个 list 存放可能创建的所有 MediaPlayer 实例，然后分别实现播放、暂停、恢复、停止等控制背景音乐播放的方法。

11.2 AudioManager

AudioManager 主要是控制音频、视频播放时的音量，还负责控制手机铃声。调整音量的方法大部分还需要指定 StreamType，声音流类型，常用的通话、系统、铃声、音乐、闹铃、通知和蓝牙等，其常用方法如表 11－3 所示。

表 11－3 AudioManager 类的常用方法

方　法	说　明
adjustStreamVolume(int streamType, int direction, int flags)	调整音量
setStreamVolume(int streamType, int index, int flags)	设置音量
getStreamVolume(int streamType)	得到音量
getStreamMaxVolume(int streamType)	得到最大音量

下面是一段增加音量的代码。

```
AudioManager am =
(AudioManager)this.getSystemService(Context.AUDIO_SERVICE);
int max = am.getStreamMaxVolume(AudioManager.STREAM_MUSIC);
int cur = am.getStreamVolume(AudioManager.STREAM_MUSIC);
if(1 + cur < =max) {
    am.setStreamVolume(AudioManager.STREAM_MUSIC, cur +1, AudioManager.FLAG
_PLAY_SOUND);
}
```

11.3 游戏中音效

为了增强效果、渲染气氛，游戏中通常会在游戏对象进行行为或产生某种结果时同时播放声音效果，例如，开枪、爆炸、碰撞甚至走路时播放现实生活中同样的音效。音效与背景音乐相比，短、小、即时，并且可能同时有多个音效需要播放，MediaPlayer 不支持多个音效同时播放。相应地，Android 定义了一个 SoundPool 类来完成音效的播放。

11.3.1 SoundPool 创建

首先要获得 SoundPool 实例，直接用 SoundPool 的构造函数来构建。

```
public SoundPool(int maxStream, int streamType, int srcQuality)
```

三个参数的含义：

- maxStream——同时播放的流的最大数量。
- streamType —— 流的类型，见 11.2 节中的 StreamType。
- srcQuality —— 采样率转化质量，现在还没效果，用 0 作为默认值。

直接使用构造函数创建在 sdk21 版本前都是这样，但之后发生改变，需要用到一个音频属性类 AudioAttributes。该类封装了音频各种属性参数，需要先设置属性，然后在构建 SoundPool 时将其传入，所以要根据运行版本设置不同创建代码。

```
public void createSoundPool()
    {
        if(Build.VERSION.SDK_INT > =Build.VERSION_CODES.LOLLIPOP)
        {
            createNewSoundPool();
        }else
        {
            createOldSoundPool();
        }




    }
    @ TargetApi(Build.VERSION_CODES.LOLLIPOP)
    void createNewSoundPool()
    {
        AudioAttributes attr = new AudioAttributes.Builder().setUsage
 (AudioAttributes.USAGE_GAME).setContentType(AudioAttributes.CONTENT_TYPE_
SONIFICATION).build();
        sp =new SoundPool.Builder().setAudioAttributes(attr).build();
    }

    @ SuppressWarnings("deprecation")
    void createOldSoundPool()
    {
        sp = new SoundPool(4,AudioManager.STREAM_MUSIC,0);
    }
 }
```

11.3.2 SoundPool 控制音效

SoundPool 加载资源有四个重载方法，分别对应不同的资源来源。

- int load(Context context, int resId, int priority)　通过资源 ID 从 APK 资源载入，这是音效常用的方法。不过音效文件放于 res 的 raw 文件夹(自行创建该文件夹)下，资源名只包括文件名，不包括其扩展名，因此，如果有一个 test. wav 文件，即不需再放一个 test. mp3 文件。

- int load(FileDescriptor fd, long offset, long length, int priority)　从 FileDescriptor 对象载入。
- int load(AssetFileDescriptor afd, int priority)　从 Asset 对象载入。
- int load(String path, int priority)　从完整文件路径名载入。

四个方法最后一个参数为优先级，目前 SDK 版本还未使用该参数，调用时可将该参数设为 1，以便和后续的 SDK 版本兼容。加载成功后，四个方法都是返回一个整形值，就是该音效的 soundID，通过该值控制音效的播放和卸载。音效一多，soundID 也就多。为了便于管理，使用 hashMap 存放这些 sound ID。有了 soundID，就可以调用 play 方法来播放。

```
int play(int soundID, float leftVolume, float rightVolume, int priority, int
loop, float rate)
```

该方法的第一个参数 soundID 指定播放哪个声音；leftVolume、rightVolume 指定左、右的音量，范围在 0 ~ 1 之间；priority 指定播放声音的优先级，数值越大，优先级越高；loop 指定是否循环，0 为不循环，-1 为一直循环；rate 指定播放的速率，范围在 0.5 ~ 2 之间，1 为正常。这个方法也返回一个整形值，播放成功是个非 0 值，称为 streamID。已经播放的音效如果要暂停、恢复、停止、设置音量、循环、优先级等都需要用 streamID，要注意不是 soundID。相应控制方法如表 11 -4 所示。

表 11 -4　音效控制方法

方　法	说　明
void pause(int streamID) ;	暂停指定音效
void resume(int streamID) ;	继续指定音效
void autoPause( )	暂停所有音效
autoResume( ) ;	继续所有音效
void stop(int streamID) ;	停止指定音效
void setVolume(int streamID, float leftVolume, float rightVolume)	设置音效左右声道音量
void setVolume(int streamID, float volume)	设置音效音量
void setPriority(int streamID, int priority)	设置优先级
void setLoop(int streamID, int loop) ;	设置循环
void setRate(int streamID, float rate) ;	设置播放速率

音效播放的简单示例代码如下：

```
HashMap<Integer, Integer> musicId = new HashMap<Integer, Integer>();
musicId.put(1,sp.load(this,R.raw.gun, 1));
musicId.put(2,sp.load(this,R.raw.cry, 1));
musicId.put(3,sp.load(this,R.raw.water,1));
int streamid = sp.play(musicId.get(1),1,1,0,0,1);
sp.stop(streamid);
```

Android 提供了一个 SparseArray 类来封装 Hash < Integer, Object > 这种结构，用起来更方便、效率更高，对应 HashMap < Integer, Integer > 就是 SparseIntArray。

```
SparseIntArray sounds = new SparseIntArray();
sounds.put(soundID, sp.load(this,R.raw.gun, 1));
int streamid = sp.play( sounds.get(soundID),1,1,0,0,1);
sp.stop(streamid);
```

12 游戏交互方式——触摸和传感器

在以往 PC 端开发游戏时，玩家与游戏交互硬件主要是键盘和鼠标，某些特殊的游戏也会添加一些专用外部设备的支持，如手柄、方向盘等等。在移动游戏开发，这些设备基本上都用不上，智能移动设备提供了全新的交互支持，包括触摸屏和各类传感器。

12.1 Touch 事件

在 Android 应用开发中，各类交互的控件、View 还有 Activity，都可以实现对 touch 事件的处理，实际上，都是响应 View 的 touch 事件。当用户手指触摸屏幕时，就会触发 touch 事件(准确地说，是触发一组 touch 事件)。手指刚接触屏幕触发 ACTION_DOWN，手指接触屏幕期间触发 ACTION_MOVE，离开屏幕时触发 ACTION_UP。

12.1.1 MotionEvent

Touch 事件封装在一个名为 MotionEvent 的类中，查看 MotionEvent 的定义，会发现除了上述三个动作，还有一堆动作定义或其他成员，实际上 MotionEvent 包括一次触摸事件全部信息、触摸坐标、手指索引、历史记录等等，其定义的触摸相关的常量如表 12-1 所示。

表 12-1 MotionEvent 中 Touch 相关常量

常量名	说 明
ACTION_DOWN	第一个手指按下
ACTION_UP	最后一个手指松开
ACTION_MOVE	移动
ACTION_CANCEL	取消
ACTION_OUTSIDE	超出边界
ACTION_POINTER_DOWN	已有手指接触屏幕情况下，新手指按下
ACTION_POINTER_UP	手指松开，但还有其他手指接触屏幕
ACTION_MASK	动作掩码，值为 0xff
ACTION_POINTER_INDEX_MASK	触点索引掩码，值为 0xff00
ACTION_POINTER_INDEX_SHIFT	8 用于移位，就是需要移动位数

之前创建的游戏视图类(无论继承自 View、SurfaceView、GLSurfaceView)中都可以重写 onTouchEvent 方法来处理 Touch 事件，代码如下：

```
public boolean onTouchEvent (MotionEvent event)
    {
        if (event. getAction () == MotionEvent. ACTION_DOWN)
        {
            float x = event. getX ();
            float y = event. getY ();
            float y2 = Utils. getScreenHeight () - y;
            Log. i (TAG,"down: x = " + x + "  y = " + y2);
        }
        if (event. getAction () == MotionEvent. ACTION_MOVE)
        {
            float x = event. getX ();
            float y = event. getY ();
            float y2 = Utils. getScreenHeight () - y;
            Log. i (TAG,"move: x = " + x + "  y = " + y2);
        }
  }
```

在上述代码中，传入的参数就是 MotionEvent 对象，通过 getAction 判断事件的动作类型，通过 getX 和 getY 分别得到 x、y 坐标，注意其坐标轴和 openGL ES 的坐标轴不同，所以对 y 值做了转换处理。

12.1.2 多点触控

在一些操作方式比较复杂的动作类游戏中，往往需要用到多点触控。多点组合加上手势，可以设计非常多的用户输入方式。对多点触控的监测，同样在 MotionEvent 中，获得动作类型的方法有两个，一个是 getAction，直接返回 mAction 值；还有一个是 getActionMasked。

```
public final int getActionMasked () {
        return mAction & ACTION_MASK;
    }
```

也就是说区别在于后者和掩码做了“与”操作，mAction 的低 8 位存放的是动作类型，高 8 位存放的则是触点的索引。通过索引可以获知该事件是由哪个触点(手指)触发的。两个掩码 ACTION_MASK 值为 0xff，ACTION_POINTER_INDEX_MASK 为 0xff00，所以很明显，和 ACTION_MASK 相“与”会屏蔽低 8 位之外的值，只留下动作类型，而和 ACTION_POINTER_INDEX_MASK 相“与”则相反，把低 8 位屏蔽，留下触点索引信息，同时再做右移 8 位的操作。表 12 - 1 中的 ACTION_POINTER_INDEX_SHIFT 值就是 8，用来做移位操作。下面是 MotionEvent 的 getActionIndex 方法实现。

```
public final int getActionIndex() {
        return (mAction & ACTION_POINTER_INDEX_MASK) >> ACTION_
POINTER_INDEX_SHIFT;
    }
```

用于多点触控的主要方法有：

- getPointerCount() 触控点的个数。
- getPointerId(int pointerIndex) pointerIndex 从 0 到 getPointerCount - 1，返回一个触摸点的标示。
- findPointerIndex(int pointerId) 返回触摸点的索引。
- getX(int pointerIndex) 通过索引得到 *X* 坐标。
- getY(int pointerIndex) 通过索引得到 *Y* 坐标。

多点触控与单点相比，就是要处理多个触控点的信息，不要混淆。每一个触点，MotionEvent 除给其分配了一个 index(索引)之外，还分配了一个 ID。从上面的方法中可以看到，大部分方法通过 index 就可以访问，例如得到第二个触点的 *X* 坐标：getX(1)。一般情况下，可能 index 就足以完成设计所需功能，可以通过循环得到所有触点的坐标，但是，在一些高级的动作设计上，并不是所有的触点都同时触发、触发之后一直存在，可能有一些触点会放开。总之在动作变化过程中，不同的事件，同一个触点的 index 可能会不同，这样不利于追踪。而 ID 只要触点没有放开，就始终保持不变，所以，如果要追踪同一个触点，可用 getPointerId 得到其 ID，通过 ID 访问其信息。由于很多方法也需要 index，因此 MotionEvent 也提供了通过 ID 的 index 方法 findPointerIndex 。

```
@Override
public boolean onTouchEvent(MotionEvent event) {
//...
 int pointerCount = event.getPointerCount();
    for (int i = 0; i < pointerCount; i++) {
     int id = event.getPointerId(i);
     int x = (int) event.getX(i);
     int y = (int) event.getY(i);

     Log.i(TAG,"Pointer" + i + ": x =" + x + " y =" + y + "id =" + id);
    }
 }
}
return true;
}
```

上述代码运行时可以从 LogCat 中观察触点的信息变化。

12.1.3 手势

单纯的一个 Action 动作，给用户提供的控制方式太少，实际上 DOWN、UP、MOVE 结合起来，可以得到很多的控制方式，再结合多点触控，就很丰富了。Android 提供了手势 GestureDetector 类来帮助开发者监测用户一些比较简单常用的手势，当然，它也是对 MotionEvent 事件进行处理。android. view. GestureDetector 类有两个回调接口 (interface)，第一个是 GestureDetector. OnDoubleTapListener，用来检测 DoubleTap 事件，就像桌面系统的鼠标的双击事件，该接口有 3 个抽象回调方法：

- onDoubleTap(MotionEvent e)　DoubleTap 双击手势事件后通知(触发)。
- onDoubleTapEvent(MotionEvent e)　DoubleTap 双击手势事件之间通知(触发)，包含 down、up 和 move 事件(这里指的是在双击之间发生的事件。例如，在同一个地方双击会产生 DoubleTap 手势，而在 DoubleTap 手势里面还会发生 down 和 up 事件，这两个事件由该函数通知)。
- onSingleTapConfirmed(MotionEvent e)　用来判定该次点击是 SingleTap 而不是 DoubleTap。如果连续点击两次就是 DoubleTap 手势；如果只点击一次，系统等待一段时间后没有收到第二次点击则判定该次点击为 SingleTap 而不是 DoubleTap，此时触发的就是 SingleTapConfirmed 事件。

第二个接口是 GestureDetector. OnGestureListener，用来通知简单常用的一些手势事件，该接口有 6 个抽象回调方法：

- onDown(MotionEvent e)　down 事件，这个就是 ACTION_DOWN，表示按下事件。
- onSingleTapUp(MotionEvent e)　一次点击 up 事件，表示按下后的抬起事件。
- onShowPress(MotionEvent e)　down 事件发生而 move 或 up 还没发生前触发该事件，此事件一般用于通知用户 press 按击事件已发生。
- onLongPress(MotionEvent e)　长按事件，down 事件后 up 事件前的一段时间间隔后(由系统分配，也可自定义)，如果仍然按住屏幕则视为长按事件。
- onFling(MotionEvent e1, MotionEvent e2, float velocityX, float velocityY)　滑动手势事件，例如 scroll 事件后突然 up，就产生 fling 事件，该事件传入的 4 个参数，e1 为起点的移动事件，e2 为当前手势点移动事件，后两者分别为滑动水平方向和垂直方向的速度。
- onScroll(MotionEvent e1, MotionEvent e2, float distanceX, float distanceY)　在屏幕上拖动事件，即 down→scroll(move)拖动→up 的移动事件。

手势类处理的也是 MotionEvent 事件，实际上手势事件的处理需要在 onTouchEvent 中调用，代码如下：

```
@ Override
    public boolean onTouchEvent(MotionEvent event) {
        mGestureDetector.onTouchEvent(event);
        return super.onTouchEvent(event);
    }

    private OnGestureListener mOnGestureListener = new OnGestureListener(){

        @ Override
        public boolean onSingleTapUp(MotionEvent e) {
            Log.i(TAG, "onSingleTapUp: " + e.toString());
            mGestureTextView.setText("onSingleTapUp: ");
            return false;

        }

        @ Override
        public void onShowPress(MotionEvent e) {
            Log.i(TAG, "onShowPress: " + e.toString());
            mGestureTextView.setText("onShowPress: ");
        }

        @ Override
        public boolean onScroll(MotionEvent e1, MotionEvent e2,
                float distanceX, float distanceY) {
            Log.i(TAG, "onScroll: "+e1.toString() + ","+e2.toString());
            mGestureTextView.setText("onScroll ");
            return false;
        }

        @ Override
        public void onLongPress(MotionEvent e) {
            Log.i(TAG, "onLongPress: " + e.toString());
            mGestureTextView.setText("onLongPress: ");
        }

        @ Override
        public boolean  onFling (MotionEvent e1,  MotionEvent  e2,  float
velocityX,
```

```
                float velocityY) {
            Log.i(TAG, "onFling: "+e1.toString() + ", "+e2.toString());
            mGestureTextView.setText("onFling ");
            return false;
        }

        @Override
        public boolean onDown(MotionEvent e) {

            Log.i(TAG, "onDown: " + e.toString());
            mGestureTextView.setText("onDown: ");

            return true;

        }
    };

    private OnDoubleTapListener mDoubleTapListener =new OnDoubleTapListener(){

        @Override
        public boolean onSingleTapConfirmed(MotionEvent e) {
            Log.i("TAG", "onSingleTapConfirmed: " + e.toString());
            mDoubleTapTextView.setText("onSingleTapConfirmed: ");
            return false;
        }

        @Override
        public boolean onDoubleTapEvent(MotionEvent e) {
            Log.i("TAG", "onDoubleTapEvent: " + e.toString());
            mDoubleTapTextView.setText("onDoubleTapEvent: ");
            return false;
        }

        @Override
        public boolean onDoubleTap(MotionEvent e) {
            Log.i("TAG", "onDoubleTap: " + e.toString());
            mDoubleTapTextView.setText("onDoubleTap: ");
            return false;
        }
    };
}
```

注意，onDown 方法处理最后返回必须是 true，而不能是 false，这和 Android 对 touch 事件处理机制有关。游戏开发相比一般应用而言，处理 touch 事件一般在一个层面即可，不会出现发送到几层 ViewGroup 的情况。前面章节阐述过 ViewGroup 的层次嵌套，touch 事件后产生，从 View 树结构中的父节点向子节点发送(DispatchTouchEvent)，如果要中断发送，可重写 onInterceptTouchEvent 拦截此次事件，如果都没有则事件发送到叶子节点，这里叶子节点的 onTouchEvent 事件可以选择处理(消费)或者发回到父节点处理，处理或不处理的依据就是 ACTION_DOWN 的返回值(因为 ACTION_DOWN 是触摸的开始动作)。如果 onTouchEvent 返回值为 true，表示处理此次事件节点，反之返回值为 false，则表示发给父节点处理，这样后续的 ACTION_MOVE、ACTION_UP 监测不到该节点。所以，在手势监测的 onDown 方法，需要返回 true，否则该事件就发给父节点处理。那么后续的其他动作该节点都不处理，监测不到也就无法组合成手势事件。

12.2 传感器

传感器系统从硬件上给智能手机添加了更丰富的功能，越来越多的开发者使用这些传感器实现更高级的应用。在游戏中的应用也不少，并且可以开发出桌面机所没有的人机互动方式。

12.2.1 Android 系统中的传感器

Android 系统支持的传感器众多，下面是常用的一些传感器信息，如表 12－2 所示。

表 12－2 常用的一些传感器信息

名　称	说　明	名　称	说　明
TYPE_ACCELEROMETER	加速度	TYPE_LIGHT	光线
TYPE_MAGNETIC_FIELD	磁场	TYPE_PRESSURE	压力
TYPE_ORIENTATION	方向	TYPE_TEMPERATURE	温度
TYPE_GYROSCOPE	陀螺仪	TYPE_PROXIMITY	接近

Android 提供的传感器数据越来越多，不止表 12－2 中的这几种，但并不是每一种传感器数据对应一个传感器硬件，有些不同的传感器数据是同一款传感器硬件做不同计算得到的，另外又有些传感器数据来源于多个传感器硬件。

这些传感器在 Android 系统以及系统自带的应用中基本上都用到。加速度传感器是最成熟的传感器，在游戏中应用非常广泛，顾名思义，是监测手机三条轴方向当前的加速度，包括重力加速度。当手机水平静止放置在桌面，只有 z 轴有加速度。手机左倾，那么 x 将变大，反之变小，这是因为有重力加速度存在。当手机不在静止状态时，例如，将手机推向右边，则手机仍然是水平的，但 x 值增加；手机做自由落体时，各方向为 0。随着版本的改进，又有了重力传感器和线性加速度传感器，前者只获得重力加速度，而后者获得除重力之外的其他加速度。

磁场传感器检测磁场，该传感器与方向传感器都是由电子罗盘传感器提供数据，指南针应用就是靠它实现的，陀螺仪返回三维方向的角加速度，旋转手机会改变三轴的数值。在3D游戏中正越来越多地被应用到。以后可能用于实现室内导航。光线传感器可以帮助调节屏幕亮度，大部分智能手机贴脸接电话时会黑屏是因为接近传感器。

12.2.2 传感器在游戏中的应用

Android系统下传感器种类众多，数据含义都不相同，但开发的基本流程相近、获取数据的过程类似、不复杂，主要有下面几个步骤：

(1)获得传感器的服务才能访问传感器，调用Context.getSystemService(SENSOR_SERVICE)方法获得SensorManager。SensorManager是所有传感器的一个综合管理类，包括传感器的种类、采样率、精准度等。

(2)通过Sensor中定义的传感器类型选择想获得的传感器。如果不确定设备具备哪些传感器，可以先通过getSensorList方法获得设备上的传感器列表，列表有搜索到的所有传感器对象，还可以通过Sensor的一组getXXX方法从对象中获得名称、类型编号、性能、版本等等。

```
List<Sensor> deviceSensors = mSensorManager.getSensorList(Sensor.TYPE_
ALL);
```

这条语句用于获得所有传感器列表，也可以改变输入参数为具体传感器类型，获得该类型的传感器列表。

(3)为传感器定义事件监听器，实现SensorEventListener接口，该接口包含两个方法：

```
public void onAccuracyChanged(Sensor arg0, int arg1)
public void onSensorChanged(SensorEvent arg0)
```

onAccuracyChanged方法在传感器精度发生变化时被调用，方法的第二个参数即为精度，精度的取值在SensorManager定义了一组常量表示：

- SensorManager.SENSOR_STATUS_ACCURACY_HIGH　报告高精度值。
- SensorManager.SENSOR_STATUS_ACCURACY_LOW　报告低精度值。
- SensorManager.SENSOR_STATUS_ACCURACY_MEDIUM　报告平均精度值。
- SensorManager.SENSOR_STATUS_ACCURACY_UNRELIABLE　报告的精度值不可靠。

传感器的精度并不是一成不变的，比如达到传感器的测量极限，或者周围环境不适合等等，这样开发者通过该方法可以监测传感器的精度改变，当精度过低，数据就可能要做一些相应的处理。当然，一般游戏应用对精度要求不会太高，所以大多数情况都不需特殊处理，做一些Log日志记录即可。

而开发用得更多的方法是onSensorChanged，该方法在传感器测量数据发生改变时调用，传递的参数包括新数据的信息。

（4）注册传感器事件监听事件。

注册传感器监听器时可以指定传感器的采样频率，可选值也通过 SensorManager 的常量定义好，有四种：

- SENSOR_DELAY_NORMAL　200 000μs，默认频率，适合响应不高的情况，例如横竖屏切换。
- SENSOR_DELAY_UI　60 000μs，适合用于 UI 界面的响应。
- SENSOR_DELAY_GAME　20 000μs，适合游戏开发。
- SENSOR_DELAY_FASTEST　0μs，最快，但大多数也没必要设置这么高，频率越高越耗电。

（5）注销传感器事件的监听。

最后一步不能少，有些传感器很耗电，不注意这一点，用户就会觉得这个应用/游戏很耗电，所以不需要时（例如 Activity 挂起）要注销事件监听，通常在 onResume 时注册监听器，而在 onPause 时注销，具体实现代码如下：

```
@ Override
protected void onCreate(Bundle savedInstanceState) {
      super.onCreate(savedInstanceState);
      requestWindowFeature(Window.FEATURE_NO_TITLE);
      getWindow().setFlags(WindowManager.LayoutParams.FLAG_FULLSCREEN,
WindowManager.LayoutParams.FLAG_FULLSCREEN);
      SetContentView(new GLGameView(this));
      SensorManager sensorManager = (SensorManager) getSystemService
(Service.SENSOR_SERVICE);
      Sensor accelerSensor = sensorManager.getDefaultSensor(Sensor.TYPE_
ACCELEROMETER);

}
public void onSensorChanged(SensorEvent event) {
      float x = event.values[0];
      float y = event.values[1];
      float z = event.values[2];

}
@ Override
public void onAccuracyChanged(Sensor sensor, int accuracy) {
}

@ Override
protected void onPause()
{
```

```
        super.onPause();
        sensorManager.unregisterListener(this);
    }
    @Override
    protected void onResume()
    {
        super.onResume();
        sensorManager.registerListener(this, accelerSensor, SensorManager.
    SENSOR_DELAY_GAME);
    }
```

12.2.3 传感器数据

onSensorChanged 中，values 的值对不同传感器有不同的含义，对于加速度传感器而言，values 变量的 3 个元素值分别表示 x、y、z 轴的加速度值。实际上，除了加速度传感器，还有重力、陀螺仪、磁场等都是代表三轴的相应值。对于移动设备而言，三条坐标轴中，x 轴与屏幕平行，是从屏幕左到右指向的坐标轴，y 轴与屏幕平行，是从下往上指向的坐标轴，而 z 则垂直屏幕，是从里向外指向的坐标轴。与 openGL 一样是右手坐标系，如图 12－1 所示。

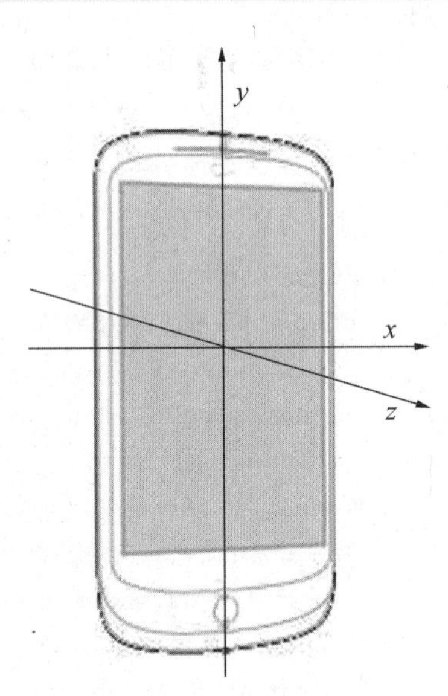

图 12－1　传感器坐标系

以加速度传感器为例，event.values 数组中的三个数据分别表示 x、y、z 轴方向的加速度减去重力加速度在该轴的分量。因此，当移动设备水平放置时，只有 z 轴有重力加速度 g。重力加速度在 z 轴的分量因为和 z 轴方向相反，即 $-g$。移动设备不动，其他加速度为 0，所以 z 轴数据就是 g。

而对于陀螺仪而言，values 数组的三个元素分别表示 x 轴旋转的角速度、y 轴旋转的角速度、z 轴旋转的角速度。手机逆时针旋转时角速度为正值，顺时针旋转时角速度为负值，所以手机水平逆时针旋转，就是绕着 z 轴转，其值为正值。

当前手机处于纵向还是横向，会影响 x、y 表示的含义，如果当前手机是纵向屏幕，那么，$x>0$ 说明当前手机左翻，$x<0$ 右翻，$y>0$ 下翻，$y<0$ 上翻。而如果当前手机是横向屏幕，那么，$x>0$ 说明当前手机下翻，$x<0$ 上翻，$y>0$ 右翻，$y<0$ 左翻。

换言之，根据横竖屏幕的不同，虽然屏幕坐标系会自动改变，但是传感器的值不会自动改变坐标系，所以这就是横屏竖屏改变时从传感器中取出的值表示的动作不一样的原因。因此游戏开发时对于人物移动、图片移动等操作，手势 x、y 正负值的含义一定要弄清楚。

一般游戏中若不允许横竖屏切换，可以在 AndroidManifest.xml 中设置 Activity 的 screenOrientation 属性。设置成 sensor，可根据传感器决定屏幕方向；设置成 nosensor 则不随传感器的数据改变。

第三篇　Android 游戏开发应用

13 搭建游戏基本框架

前面章节阐述了游戏开发中涉及的关键技术，本章应用这些知识来定义游戏开发所必需的基本类。

13.1 图形渲染

在项目中创建 view 包来存放图形渲染相关的类，该包包括 VertexData、Frame、Sprite、SpriteButton、ResManager 和 GameGLView 六个类。

13.1.1 VertexData 顶点数据类

由于项目采用 OpenGl ES1. x 来进行图形渲染，因此定义 VertexData 用来对顶点数据进行封装，包括顶点坐标、颜色和纹理坐标均使用该类。其中包括一个浮点型数组用来保存顶点原始数据，而另一个成员变量浮点型缓冲用来保存转换后数据存放的缓冲。除了构造函数外，实现两个方法分别用于更新缓冲，主要针对顶点数据动态改变，缓冲需要重写的情况，以及 bufferUtil 用于将浮点型数组数据填充入浮点型缓冲中。代码如下：

```
public class VertexData {
    public FloatBuffer buffer;
    public float[] vertices;
    public VertexData(float[] vertices)
    {
        this.vertices = vertices;
        buffer = bufferUtil(vertices);
    }
    public void updateVertex()
    {
        buffer = bufferUtil(vertices);
    }
    private FloatBuffer bufferUtil(float []arr)
    {
        FloatBuffer buffer;
        ByteBuffer vbb = ByteBuffer.allocateDirect(arr.length * 4);
```

```
            vbb.order(ByteOrder.nativeOrder());
            buffer = vbb.asFloatBuffer();
            buffer.put(arr);
            buffer.position(0);
            return buffer;
        }
}
```

13.1.2 Frame 类图像的帧信息类

在图形渲染一章中，介绍了如何使用纹理和四边形来显示一幅图像。实际上游戏中的对象为了更好的效果，多数并非静态的图像，而是由一组帧图像顺序播放从而组成动画，组成动画帧的各个图像并非一定是独立的图像文件。为了提高效率，通常游戏项目会把需要用到的静态帧图像处理在同一张图片上，如图 13－1 所示。

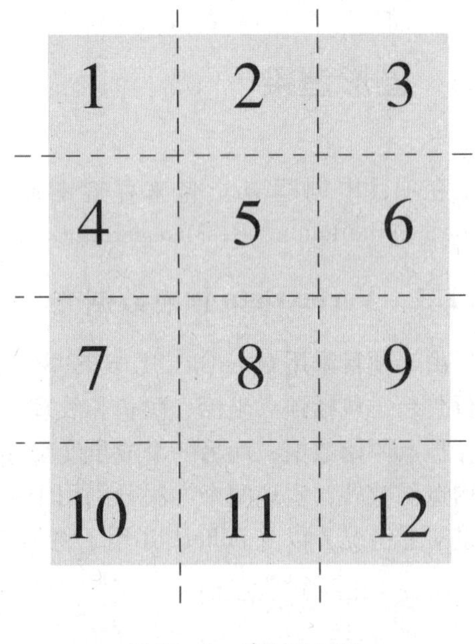

图 13－1　素材动画图

图 13－1 中包括了 1—12 共 12 个数字的图像，如果要将图中的数字图像组成一组动画播放，就需要对该图片进行分割。在 OpenGl ES 中，可以通过指定纹理坐标来实现，例如，显示数字 1 对应四边形四个点左上、右上、左下、右下的纹理坐标分别为（0，0）、（0.33，0）、（0，0.25）、（0.3，0.25），而每一个其余的数字也都有相应的一组纹理坐标来表示。Frame 这个类就用于保存指定纹理中的相应坐标，从而向 Sprite 提供纹理中的部分图像的位置。因此，Frame 类中有两个成员变量，纹理索引和 VertexData 对象，前者用于指定使用哪张纹理，后者则保存纹理坐标，即目标纹理的裁剪位置。类定义如下：

```
public class Frame {
     public Integer texid;
     public VertexData texcoord;

     public Frame()
     {
          texcoord = new VertexData(new float[]{
                    0,0,
                    0,1,
```

```
                1,0,
                1,1,
        });
        texid = -1;
    }

    public Frame(ResManager resManager,int resid)
    {
        texcoord = new VertexData(new float[]{
                0,0,
                0,1,
                1,0,
                1,1,
        });
        if(resManager.getBmpSize(resid)==null)
        {

            texid = -1;
            return;
        }
        this.texid = resManager.getTexID(resid);
    }
    public Frame(ResManager resManager,int resid,float x,float y,float
width,float height)
{
        if(resManager.getBmpSize(resid)==null)
        {
            texcoord = new VertexData(new float[]{
                0,0,
                0,1,
                1,0,
                1,1,
            });
            texid = -1;
            return;
        }
        //像素单位转为纹理坐标
        x /= resManager.getBmpSize(resid).width;
        y /= resManager.getBmpSize(resid).height;

        width /= resManager.getBmpSize(resid).width;
```

```
        height /= resManager.getBmpSize(resid).height;

        texcoord = new VertexData(new float[]{

                x,y,
                x,y+height,
                x+width,y,
                x+width,y+height,
        });

        this.texid = resManager.getTexID(resid);
    }
    //用于从一张纹理中生成一组frame对象,例如图13-1中,可以得到12个frame对象,
//将其放置Frame数组中
    public static Frame[] generate(ResManager resManager,int resid,float
x,float y,float width,float height,Integer row,Integer col,Integer number)
    {
        Frame[] frames = new Frame[number];
        int index = 0;
        for (int i = 0;i<row;i++)
        {
            for(int j=0;j<col;j++)
            {
                frames[index]=new
Frame(resManager,resid,x+j* width,y+i* height,width,height);
                index++;
                if(index == number)
                    return frames;
            }
        }
        return frames;
    }
}
```

这样,一个Frame对象就对应要渲染一帧图像的位置信息。一个动画就需要一个frame数组来保存逐帧的图像位置信息。

13.1.3 Sprite 精灵类

大部分的游戏引擎在实现渲染模块时，会有个 Sprite 精灵类，可见对象大多是基于精灵类来绘制的。有一些引擎用 Sprite 表示单帧精灵，有些则表示动画精灵。在本项目中，两者都可以通过 Sprite 来绘制。前面定义的 Frame 类保存了纹理索引及纹理中的裁剪位置，要渲染图像必然要使用到 Frame。事实上，Sprite 类中，有一个 Frame 数组，用于记录该精灵的动画帧的纹理信息。两个类以及 VertexData 类关系如图 13－2 所示。

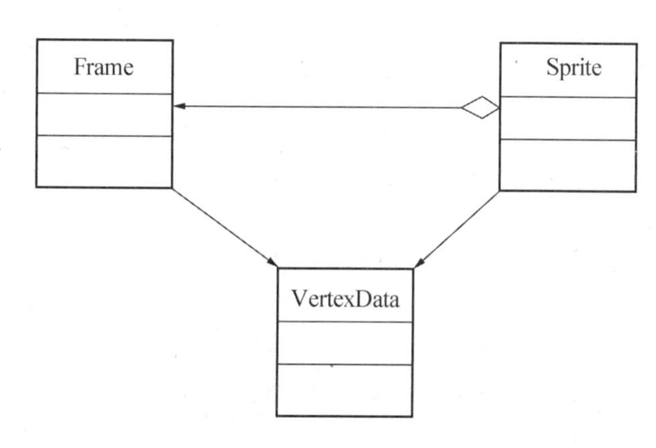

图 13－2　Frame、Sprite、VertexData 类关系图

Sprite 除了需要 Frame 的纹理坐标之外，还需要顶点的位置和颜色数据，Sprite 创建四边形，再将纹理映射到四边形从而实现位图的绘制。指定颜色的作用有两个，一个是当纹理数据不存在，或者想渲染没有纹理的四边形时，可以指定颜色；另一个就是将颜色数值与纹理相乘，从而改变精灵的色调。这一点，需要通过纹理混合来实现。限于篇幅，这里不详细介绍纹理混合。混合方式默认是 Replace，当有纹理时，颜色数据不起作用，为了实现混合相乘效果，需要改变混合方式，其语句如下：

```
GLES10.glTexEnvf(GLES10.GL_TEXTURE_ENV, GLES10.GL_TEXTURE_ENV_MODE,
GLES10.GL_MODULATE);
```

对于 Sprite 而言，如果要显示动画的播放，那么还需要对帧更进一步管理，包括设置帧播放间隔时间（播放速度）、暂停、播放、设置循环等，最关键的方法在于 next，即播放下一帧，其处理流程如图 13－3 所示。

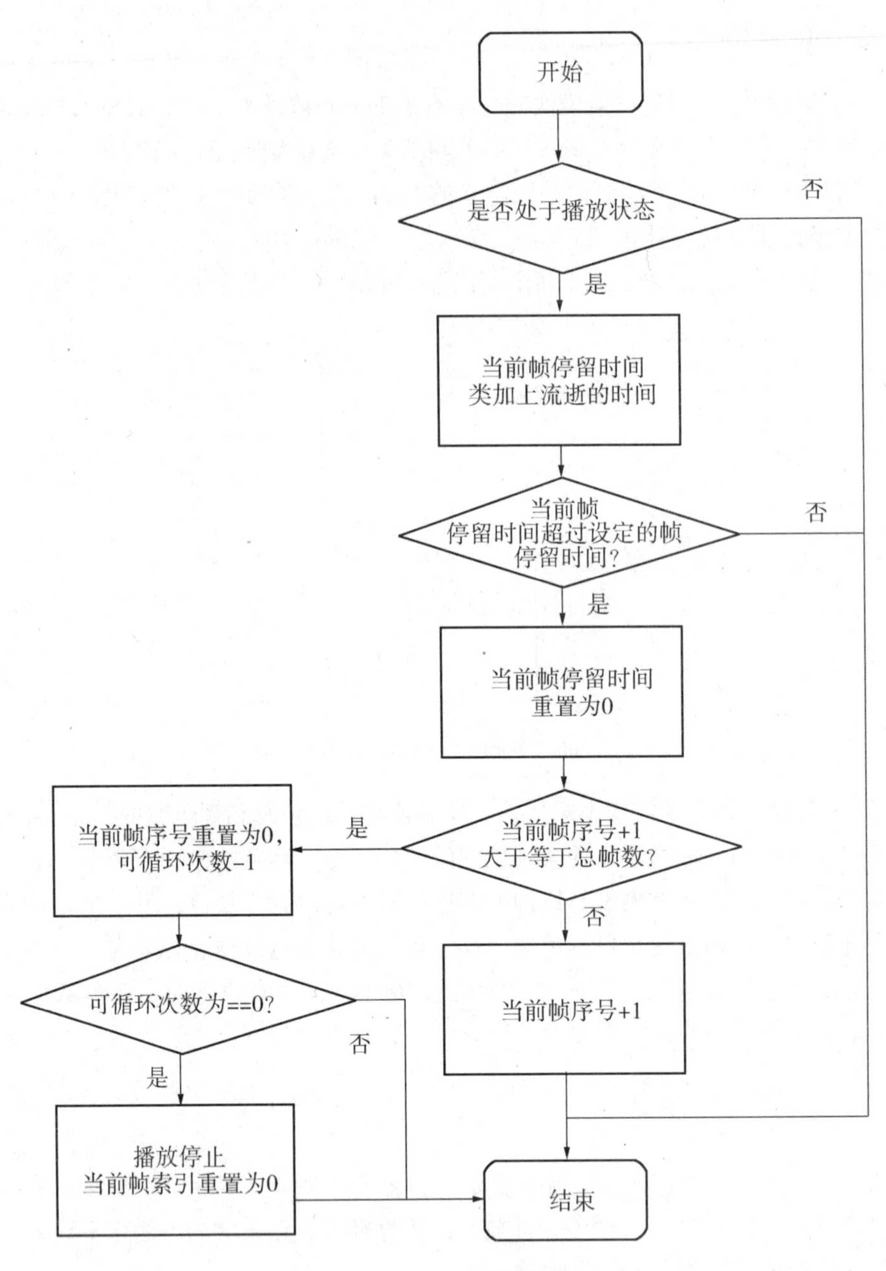

图 13－3　帧播放流程

该类实现的关键代码如下：

```
public class Sprite {
    private float posx;
    private float posy;
    private float width;
    private float height;
```

```
private float angle;
private boolean visible = true;
private ArrayList<Frame> frames;
private int currentFrameIndex = 0;
private boolean bPlay = true;
private boolean bPause = false;
private int loopCount = 0 ;
private long intervalTime = 0;
private long curFrameStayTime = 0;
private float[] colors;

private Rect spriteRect = new Rect();

private static VertexData position = new VertexData(new float[]{
          -0.5f,0.5f,0.0f,
          -0.5f, -0.5f,0.0f,
          0.5f,0.5f,0.0f,
          0.5f, -0.5f,0.0f
});
private VertexData color = new VertexData(new float[]{
          1.0f,1.0f,1.0f,1.0f,
          1.0f,1.0f,1.0f,1.0f,
          1.0f,1.0f,1.0f,1.0f,
          1.0f,1.0f,1.0f,1.0f
});

public Sprite(float x,float y,float width,float height,Frame[] frames)
{
     this.posx = x;
     this.posy = y;
     this.width = width;
     this.height = height;
     colors = new float[]{1.0f,1.0f,1.0f,1.0f};
     if(frames!=null)
            this.frames = new ArrayList<Frame>(Arrays.asList
            (frames));
     else
            this.frames = new ArrayList<Frame>();
     curFrameStayTime = 0;
     loopCount = -1;
```

```
        intervalTime = 300;
        angle = 0;

spriteRect.set(Math.round(x - width/2),Math.round(y - height/2),Math.round
(x + width/2),Math.round(y + height/2));
    }
    public Sprite(float x,float y,float width,float height)
    {
        this(x, y, width, height, null);
    }

    public Sprite(float width,float height)
    {
        this(0, 0, width, height, null);
    }
    public void add(Frame frame)
    {
        frames.add(frame);
    }
    public void add(Frame[] frameArray)
    {
        for(int i =0;i < frameArray.length;i + +){
            this.frames.add(frameArray[i]);
        }
    }
    public void changeFrames(Frame[] frames){
        this.frames = new ArrayList < Frame > (Arrays.asList(frames));
    }
    public void setColor(float red,float green,float blue,float alpha)
    {
        colors[0] =red;
        colors[1] =green;
        colors[2] =blue;
        colors[3] =alpha;
        System.arraycopy(colors,0,color.vertices,0,4);
        System.arraycopy(colors,0,color.vertices,4,4);
        System.arraycopy(colors,0,color.vertices,8,4);
        System.arraycopy(colors,0,color.vertices,12,4);
        color.updateVertex();
    }
```

```
    public void draw()//,Scale scale)
    {
        if(!visible)
            return;
        boolean bTex = true;
        if(frames.size() == 0 || frames.get(currentFrameIndex) == null ||
frames.get(currentFrameIndex).texid == -1)
        {
            bTex = false;
        }
        GLES10.glPushMatrix();
        GLES10.glTranslatef(posx, posy, 0);
        GLES10.glScalef(width, height, 0);
        GLES10.glEnableClientState(GLES10.GL_VERTEX_ARRAY);
        GLES10.glVertexPointer(3, GLES10.GL_FLOAT, 0, position.buffer);

        if(bTex)
        {
GLES10.glTexEnvf(GLES10.GL_TEXTURE_ENV, GLES10.GL_TEXTURE_ENV_MODE,
GLES10.GL_MODULATE);
            GLES10.glEnableClientState(GL10.GL_COLOR_ARRAY);
            GLES10.glColorPointer(4, GL10.GL_FLOAT, 0, color.buffer);

            GLES10.glColor4f(colors[0], colors[1], colors[2], colors[3]);

            GLES10.glEnableClientState(GL10.GL_TEXTURE_COORD_ARRAY);
            GLES10.glTexCoordPointer(2, GL10.GL_FLOAT, 0, frames.get
(currentFrameIndex).texcoord.buffer);
            GLES10.glBindTexture(GL10.GL_TEXTURE_2D, frames.get
(currentFrameIndex).texid);
            GLES10.glTexParameterf(GL10.GL_TEXTURE_2D, GL10.GL_
TEXTURE_MIN_FILTER, GL10.GL_LINEAR);
            // GL10.GL_LINEAR_MIPMAP_NEAREST);
            GLES10.glTexParameterf(GL10.GL_TEXTURE_2D, GL10.GL_TEXTURE_MAG_
FILTER, GL10.GL_LINEAR);//_MIPMAP_LINEAR);
            GLES10.glEnable(GL10.GL_TEXTURE_2D);
        }else
        {
            GLES10.glEnableClientState(GL10.GL_COLOR_ARRAY);
```

```
        GLES10.glColorPointer(4, GL10.GL_FLOAT, 0, color.buffer);

        GLES10.glColor4f(colors[0], colors[1], colors[2], colors[3]);
    }

        GLES10.glEnable(GL10.GL_ALPHA_TEST);
        GLES10.glAlphaFunc(GL10.GL_GREATER, 0.1f);
        GLES10.glFrontFace(GL10.GL_CW);
        GLES10.glDrawArrays(GL10.GL_TRIANGLE_STRIP, 0, 4);
        if(bTex) {
            GLES10.glDisableClientState(GL10.GL_TEXTURE_COORD_ARRAY);
            GLES10.glDisable(GL10.GL_TEXTURE_2D);

            GLES10.glDisableClientState(GL10.GL_COLOR_ARRAY);
        }else
        {
            GLES10.glDisableClientState(GL10.GL_COLOR_ARRAY);
            }
            GLES10.glDisableClientState(GL10.GL_VERTEX_ARRAY);

            GLES10.glPopMatrix();
    }

    public void next(long interval)
    {
        if(!bPlay)
            return;
        curFrameStayTime += interval;
        if(curFrameStayTime < intervalTime)
        {
            return;
        }
        curFrameStayTime = 0;
        //下面代码在并非原子操作,在多线程,注意不能让 currentFrameIndex >=
//frames.size()

        if((currentFrameIndex+1) >=frames.size())
        {
            currentFrameIndex = 0;
            if(--loopCount==0)
            {
```

```
                bPlay = false;
                currentFrameIndex = 0;
            }
        }else
        {
            currentFrameIndex ++;
        }
    }

    public void stop()
    {
        bPlay = false;
        currentFrameIndex = 0;
    }

    public void pause()
    {
        bPlay = false;
        bPause = true;
    }

    public void resume()
    {
        bPlay = true;
    }

    public void play()
    {
        bPlay = true;
        currentFrameIndex = 0;
    }
}
```

13.1.4 SpriteButton 精灵按钮类

在游戏中，为了美观，也为了与其他的可绘制对象统一管理，通常不会直接使用Android 提供的控件，而是自行绘制和处理相应的事件。在本项目中，只需要用按钮来作为进入菜单，因此自定义一个精灵按钮类来实现这一功能。该类功能比较简单，用两个精灵对象成员分别表示按钮的 normal（普通）状态和 touched（被点击）状态，然后处理传入的 touch 事件。代码如下：

```
public class SpriteButton {
    private Sprite normal;
    private Sprite touched;
    private GameState state;
    private int index;
    public SpriteButton (GameState state, int index, Sprite normal, Sprite
touched)
    {
        this.state = state;
        this.index = index;
        this.normal = normal;
        this.touched = touched;
        reset();
    }

    public void reset()
    {
        normal.setVisible(true);
        touched.setVisible(false);
    }

    public void setTouched()
    {
        normal.setVisible(false);
        touched.setVisible(true);
    }

    public boolean isNormalState()
    {
        return normal.isVisible();
    }

    public void handleEvent(MotionEvent event)
    {
        if(event.getAction() == MotionEvent.ACTION_DOWN)
        {
            int x = (int)event.getX();
            int y = Config.getInstance().screenSize.y - (int)event.getY
();

            if(normal.isCollision(new Point(x,y)) && isNormalState())
            {
                setTouched();
            }
```

```
            }

            if(event.getAction() == MotionEvent.ACTION_UP)
            {
                int x = (int)event.getX();
                int y =Config.getInstance().screenSize.y - (int)event.getY();

                if(normal.isCollision(new Point(x,y)) && !isNormalState())
                {
                    state.handleBtnEvent(index);
                }
                reset();
            }
        }
    }
```

13.1.5 ResManager 纹理资源管理类

纹理的加载已在7.5节述及，需要先通过位图工厂加载位图资源，然后再生成纹理，将位图绑定至纹理上。该类比较简单，负责加载位图，生成纹理，以及为其他类提供纹理索引。注意，该类维护了一个HashMap，用于存放诸多纹理资源，而采用的key值则是在R文件中的资源ID。这样获取纹理索引只需要提供对应的资源ID即可。为了便于资源的加载，本类设计为单例。关键代码如下：

```
public class ResManager {
    private Map<Integer,Bitmap> bitmaps = new ConcurrentHashMap
<Integer,Bitmap>();
    private Map<Integer,Size> bmpSizes = new HashMap<Integer,Size>();

    private SparseIntArray texs = new SparseIntArray();

    private Resources res;

    private static ResManager instance = null;
    private ResManager() {}

    public void releaseTexRes()
    {
        texs.clear();
```

```
    }
    public static ResManager getInstance()
    {
        if(instance == null)
        {
            synchronized (ResManager.class){
                if(instance == null)
                    instance = new ResManager();
            }
        }
        return instance;
    }

    public void setResource(Resources res)
    {
        this.res = res;
    }

    public Bitmap loadBitmap(int resid)
    {
        Bitmap bmp = null;
        if(bitmaps.get(resid) ==null) {
            BitmapFactory.Options options =new BitmapFactory.Options();
            options.inScaled = false;
            bmp = BitmapFactory.decodeResource(res, resid);
            bitmaps.put(resid, bmp );
        }
         return bmp;
    }

    public int genTexture(int resid) {
        if(texs.get(resid)! =0)
            return -1;
        Bitmap bmp = bitmaps.get(resid);
        if (bmp == null)
        {
            bmp = loadBitmap(resid);
        }

        if (bmp == null)
```

```
            return -1;

        bmpSizes.put(resid, new Size(bmp.getWidth(), bmp.getHeight()));

        int[] temp = new int[1];
        GLES10.glGenTextures(1, temp, 0);
        int id = temp[0];

        texs.put(resid, id);
        GLES10.glBindTexture(GL10.GL_TEXTURE_2D, id);

        GLES10.glTexParameterf(GL10.GL_TEXTURE_2D, GL10.GL_TEXTURE_
MIN_FILTER, GL10.GL_LINEAR);
        GLES10.glTexParameterf(GL10.GL_TEXTURE_2D, GL10.GL_TEXTURE_
MAG_FILTER, GL10.GL_LINEAR);
        GLES10.glTexParameterf(GL10.GL_TEXTURE_2D, GL10.GL_TEXTURE_
WRAP_S, GL10.GL_CLAMP_TO_EDGE);
        GLES10.glTexParameterf(GL10.GL_TEXTURE_2D, GL10.GL_TEXTURE_
WRAP_T, GL10.GL_CLAMP_TO_EDGE);
        GLES10.glTexEnvf(GL10.GL_TEXTURE_ENV, GL10.GL_TEXTURE_ENV_
MODE, GL10.GL_REPLACE);
        GLUtils.texImage2D(GL10.GL_TEXTURE_2D, 0, bmp, 0);
        GLES10.glTexParameterx(GL10.GL_TEXTURE_2D, GL10.GL_TEXTURE_
MIN_FILTER, GL10.GL_LINEAR);
        GLES10.glTexParameterx(GL10.GL_TEXTURE_2D, GL10.GL_TEXTURE_
MAG_FILTER, GL10.GL_LINEAR);
        return id;
    }


    public Size getBmpSize(int resID)
    {
        return bmpSizes.get(resID);
    }

    public int getTexID(int resID)
    {
        int temp = texs.get(resID, -1);

        return temp;
    }
}
```

13.1.6 GameGLView 游戏视图类

GameGLView 是游戏渲染的主类，在该类中，创建了渲染线程(父类 GLSurfaceView 创建)，该类大部分代码在第6章已有阐述，此节重点介绍该类的执行流程，以及各个阶段应该处理的工作。该类中定义的内部类 GameRenderer 渲染器实现了 surfaceCreated、surfaceChanged、onDrawFrame 方法，其中 surfaceCreated 在表面创建时会被调用，而 surfaceChanged 在表面改变时会调用。实际上，当表面创建完成即会调用一次 surfaceChanged，而 onDrawFrame 是通过渲染线程来循环进行帧的绘制的。正常情况下其执行顺序为 Created→Changed→onDrawFrame，但对于移动设备而言，接听电话、收看短信等导致游戏中断暂停是非常普遍的，在 4.1.2 节“Activity 生命周期”中已经介绍过关于各种情况的处理，而当游戏被切入后台时，不能持续维持帧的渲染，所以需要在 Activity 的 onResume 和 onPause 中分别对渲染线程进行恢复和暂停。代码如下：

```
@ Override
protected void onPause()
{
    super.onPause();
    view.onPause();
    if(config.sensor)
    {
        sensorManager.unregisterListener(this);
    }
}
@ Override
protected void onResume()
{
    super.onResume();
    view.onResume();
    if(config.sensor)
    {
        sensorManager.registerListener(this, accelerSensor,
SensorManager.SENSOR_DELAY_GAME);
    }
}
```

从上述代码可见，在 Activity 的暂停和恢复中，对渲染线程同样进行暂停恢复，当然，还有传感器监听也应该在应用暂停时暂停。这样，通过上面代码，就可以避免渲染线程做无谓的工作。但随之而来的问题是，当 Activity 恢复时会发生纹理全部丢失的情况。实际上，加载的 Bitmap 位图资源还在，但绑定的纹理不见了，需要重新加载。而无论是 Activity 暂时切除然后回来还是完全销毁再重新打开，onSurfaceCreated 都会重新调用，所以只需在该方法中重新生成纹理即可。代码如下：

```
@ Override
public void onSurfaceCreated(GL10 gl, EGLConfig config) {
      GLES10.glClearColor(0.1f, 0.1f, 0.1f, 1.0f);
      GLES10.glShadeModel(GL10.GL_SMOOTH);
      GLES10.glClearDepthf(1.0f);
      GLES10.glEnable(GL10.GL_DEPTH_TEST);
      GLES10.glDepthFunc(GL10.GL_LEQUAL);
      GLES10.glBlendFunc(GL10.GL_SRC_ALPHA, GL10.GL_ONE_MINUS_SRC_
ALPHA);
      GLES10.glEnable(GL10.GL_BLEND);
      GLES10.glHint (GL10.GL_PERSPECTIVE_CORRECTION_HINT, GL10.GL_
NICEST);

      ResManager.getInstance().releaseTexRes();
      ResManager.getInstance().genTexture(R.drawable.welcome);
      ResManager.getInstance().genTexture(R.drawable.plane);
}
```

注意：重新生成纹理之前需要将资源管理中原本的 hashMap 清空。

13.2 音乐播放

第11章介绍了游戏背景音乐和音效的处理方法。在本节中，定义 GameAudio 来管理所有声音，代码如下：

```
public class GameAudio {
      private Map<Integer,MediaPlayer> musics = new HashMap
<Integer,MediaPlayer>();
      private List<MediaPlayer> playingMusics = new ArrayList
<MediaPlayer>();
      private SoundPool soundPool;
      private SparseIntArray sounds = new SparseIntArray();

      private Context context;

      public GameAudio(Context context)
      {
            this.context = context;
            init();
      }
      public void init()
```

```
    {
        createSoundPool();
    }

    private void createSoundPool()
    {
        if(Build.VERSION.SDK_INT>=Build.VERSION_CODES.LOLLIPOP)
        {
            createNewSoundPool();
        }else
        {
            createOldSoundPool();
        }
    }
    @TargetApi(Build.VERSION_CODES.LOLLIPOP)
    private void createNewSoundPool()
    {
        AudioAttributes attr = new AudioAttributes.Builder().setUsage
(AudioAttributes.USAGE_GAME).setContentType(AudioAttributes.CONTENT_TYPE_
SONIFICATION).build();
        soundPool = new SoundPool.Builder().setAudioAttributes(attr)
.build();
    }

    @SuppressWarnings("deprecation")
    private void createOldSoundPool()
    {
        soundPool = new SoundPool(4, AudioManager.STREAM_MUSIC,0);
    }

    public void loadSound(int resid)
    {
    //将资源ID作为音效列表中的key
        sounds.put(resid,soundPool.load(context,resid,1));
    }

    public int playSound(int soundresID)
    {
        int key = sounds.get(soundresID);
        if(key==0)
            return 0;
```

```
        return soundPool.play(key, 1, 1, 0, 0,1);
}

    public void stopSound(int streamID)
    {
        soundPool.stop(streamID);
    }

    public void pauseSound(int streamID)
    {
        soundPool.pause(streamID);
    }

    public void resumeSound(int streamID)
    {
        soundPool.resume(streamID);
    }

    public void loadMusic(int resid)
    {
        musics.put(resid, MediaPlayer.create(context, resid));
    }
    public void playMusic(int resmusicid)
    {
        MediaPlayer mp = musics.get(resmusicid);
        if(mp == null)
            return;

        if(mp.isPlaying())
        {
            mp.pause();
            mp.seekTo(0);
        }else {
            mp.setOnErrorListener(new MediaPlayer.OnErrorListener() {
                @ Override
                public boolean onError (MediaPlayer mp, int what, int
extra) {
                    mp.reset();
                    return false;
                }
            });
        }
```

```
        mp.start();
        playingMusics.add(mp);
    }
    public void pauseMusic(int resmusicid)
    {
        MediaPlayer mp = musics.get(resmusicid);
        if(mp == null)
            return;
        mp.pause();
    }
    public void stopMusic(int resmusicid)
    {
        MediaPlayer mp = musics.get(resmusicid);
        if(mp == null)
            return;
        mp.stop();
        playingMusics.remove(mp);
    }
    public void stopAllMusic()
    {
        Iterator<MediaPlayer> iter = playingMusics.iterator();
        while(iter.hasNext())
        {
            iter.next().stop();
        }
    }

    public void resumeMusic(int resmusicid)
    {
        MediaPlayer mp = musics.get(resmusicid);
        if(mp == null)
            return;
        mp.start();
    }
}
```

13.3 数据存储加载

游戏中有些配置的参数需要保存，例如是否开启声音、传感器、网络，这里用一个配置类来保存这些信息。该类为单例，除了一些配置参数，屏幕尺寸也保存在这里，代码如下：

```
public class Config {
    static public boolean sensor = false;
    static public boolean music = false;
    static public boolean sound = false;
    static public boolean text = true;
    static public boolean net = false;
    static public Point screenSize = new Point();
    private static Config instance = null;
    private Config() {}

    public static Config getInstance()
    {
        if(instance == null)
        {
            synchronized (Config.class){
                if(instance == null)
                    instance = new Config();
            }
        }
        return instance;
    }
}
```

在 Activity 创建时，读取 SharedPreference 为 config 赋值。

```
config = Config.getInstance();
getWindowManager().getDefaultDisplay().getSize(config.screenSize);
sp = this.getSharedPreferences(GameConst.DataStorage.SP_NAME,
Context.MODE_PRIVATE);
Config.net = sp.getBoolean("net", false);
Config.music = sp.getBoolean("music", false);
Config.sound = sp.getBoolean("sound",false);
Config.sensor = sp.getBoolean("sensor",false);
Config.text = sp.getBoolean("text",false);
```

游戏的逻辑相关的数据保存在 Sqlite 中。在 8.3 节中定义了合同类和 SQliteOpenHelper 的子类来简单处理。数据库中有几个表，就定义几个对应的合同类，保存表名和列信息，在 SQLiteOpenHelper 类中完成对具体数据读写的方法。

13.4 网络通信与多线程

13.4.1 网络通信

网络通信方面一共包括三个类，即 NetProxy、RingBuffer、DataReader，已经在第 10 章介绍过。本节介绍 Activity 如何启动网络线程。就网络连接监听线程而言，由于 NetProxy 继承自 Thread，因此直接构建实例，然后 start 即可。而从环形缓冲中读取数据的 DataReader 实现了 Runnable 接口，所以定义了一个 HandlerThread 线程，当需要读取数据时，将 DataReader 实例 post 给该线程工作。代码如下：

```
if(config.net)
{
    readFromRingThread = new HandlerThread("readFromRing");
    readFromRingThread.start();
    readHandler = new Handler(readFromRingThread.getLooper());

    ringBuffer = new RingBuffer(65535);
    NetProxy netProxy = new NetProxy(mainHanler,ringBuffer);
    netProxy.start();
}
```

而当收到网络数据消息事件时，去读取环形缓冲数据。

```
public void readData()
{
    readHandler.post(new DataReader(mainHanler,ringBuffer));
}
```

从上述代码可以看到，我们将读取数据线程的 Handler 保存在 Activity 中，当读取数据并解析完成后，提交给 Activity 来处理。

13.4.2 多线程

实际上，由于项目比较简单，并未为游戏逻辑处理开辟一个线程。如果逻辑比较复杂，代码较多，可能需要单独开线程处理。由于 Activity 接收消息较多，为其定义一个回调类来完成消息的处理工作。

```
public class MainCallback implements Handler.Callback {
    private PlaneGameActivity activity;

    public MainCallback(PlaneGameActivity activity)
    {
        this.activity = activity;
    }

    @Override
    public boolean handleMessage(Message message) {
        switch (message.what)
        {
            case GameConst.HandleMsg.CONN_SUCCESS:
                break;
            case GameConst.HandleMsg.RECV_DATA:
                activity.readData();
                break;
            case GameConst.HandleMsg.NEW_PLAYER:
                break;
            case GameConst.HandleMsg.SUR_CHANGE:
                break;
            case GameConst.HandleMsg.BEGIN_FRAME_DRAW:
                activity.updateLogic((long)message.obj);
                break;
        }
        return false;
    }
}
```

而在 Activity 中，创建该类实例，并将其传递给需要向主线程发送消息的子线程。例如 readFromRingThread 线程。

```
mainHandler = new Handler(new MainCallback(this));
```

13.5 场景状态管理

一个完整的游戏会包含很多的状态/场景，如欢迎和加载、主菜单、游戏主场景、游戏结束，或排行榜、商店等等。简单的处理是，使用状态模式，而如果状态之间的关系复杂，且跳转较频繁，则需要使用有限状态机。在本项目中，状态比较简单，且彼此关系单一，在这种情况下，只需定义两个类：状态基类和状态管理类。

13.5.1 GameState状态基类

无论处于哪个状态，游戏都需要渲染该状态显示的画面，而画面由精灵组成，所以在基类中，维护一个list数组，保存该状态下的精灵。为了保证每个精灵都在精灵列表中，应用工厂模式，提供createSprite方法来创建精灵，创建实例后直接加入列表。为了处理状态之间跳转的情况，分别定义四个方法：onEnter（进入）、onExit（退出）、onLeave（离开）、onBack（回来）。该类代码如下：

```
public abstract class GameState {

    protected Point screenSize = Config.getInstance().screenSize;
    protected List<Sprite> sprites = new ArrayList<Sprite>();

    public Sprite createSprite(float x,float y,float width,float height,
Frame[] frames)
    {
            Sprite sprite = new Sprite(x, y, width, height, frames);
        synchronized (sprites) {
            sprites.add(sprite);
        }
        return sprite;
    }

    public Sprite createSprite(float x,float y,float width,float height)
    {
        return createSprite(x, y, width, height, null);
    }

    public Sprite createSprite(float width,float height)
    {
        return createSprite(0,0,width,height);
    }
    public void onEnter()
    {
    }
    public void onExit()
    {
    }
    public abstract void onBack();
```

```
    public abstract void onLeave();
    public void draw()
    {
        synchronized (sprites) {
            for (Sprite sp : sprites) {
                sp.draw();
            }
        }
    }
    public  void update(long interval,GameStateManager gsm)
    {
        synchronized (sprites) {
            for (Sprite sp : sprites) {
                sp.next(interval);
            }
        }
    }
    public abstract void handleEvent(MotionEvent event);
    public void handleBtnEvent(int btnIndex){}
    public boolean isNeverEnter()
    {
        return bNeverEnter;
    }
    public abstract boolean isEnded();
}
```

定义了状态基类，根据游戏项目实际的情况来定义其子类，分别对应游戏中具体的状态。有了各个状态之后，再定义状态管理类来协调和转换状态。

13.5.2 GameStateManager 状态管理类

状态管理类有三个功能：新加状态、转换状态、管理状态。在该类中定义枚举类型来包括所有的状态类型，然后包含一个 HashMap 保存所有状态，键值就是枚举类型，保存当前状态，并接受逻辑线程和渲染线程通知，调用当前状态的更新和渲染方法。代码如下：

```
public class GameStateManager {
    public enum STATE
    {
        LOADING,MENU,
        GAMEMAIN,GAMEOVER
    }
```

```
Map < STATE,GameState > states = new HashMap < STATE,GameState > ();
GameState currentState;
STATE currentStateName;

public void addState(STATE name,GameState state)
{
    if(states.containsKey(name))
        return;
    states.put(name,state);
}
public void setState(STATE name)
{
    if(name == currentStateName)
    {
        return;
    }
    if(states.containsKey(name))
    {
        if(currentState! =null)
            currentState.onLeave();
        if(states.get(name).isNeverEnter())
            states.get(name).onEnter();
        else
            states.get(name).onBack();
        setCurrentState(name);
    }else
    {
        //Log.i("STATE","State no find!");
        return;
    }
}

private void setCurrentState(STATE name)
{
    currentStateName = name;
    currentState = states.get(name);
}

public void draw()
{
    if(currentState! =null)
        currentState.draw();
}

public void update(long interval)
{
    currentState.update(interval,this);
```

```
    }

    public void handleEvent(MotionEvent event)
    {
        currentState.handleEvent(event);
    }

}
public class GameStateManager {
    public enum STATE
    {
        LOADING,MENU,
        GAMEMAIN,GAMEOVER
    }

    Map<STATE,GameState> states=new HashMap<STATE,GameState>();
    GameState currentState;
    STATE currentStateName;

    public void addState(STATE name,GameState state)
    {
        if(states.containsKey(name))
            return;
        states.put(name,state);
    }
    public void setState(STATE name)
    {
        if(name==currentStateName)
        {
            return;
        }
        if(states.containsKey(name))
        {
            if(currentState!=null)
                currentState.onLeave();
            if(states.get(name).isNeverEnter())
                states.get(name).onEnter();
            else
                states.get(name).onBack();
            setCurrentState(name);
        }else
        {
            return;
        }
    }

    private void setCurrentState(STATE name)
    {
```

```
        currentStateName = name;
        currentState = states.get(name);
    }

    public void draw()
    {
        if(currentState != null)
            currentState.draw();
    }

    public void update(long interval)
    {
        currentState.update(interval,this);
    }

    public void handleEvent(MotionEvent event)
    {
        currentState.handleEvent(event);
    }

}
```

13.6 工具及其他类

游戏中除了上面常用模块的基础类，还定义了一组相关辅助类，如字节转换、常量、向量类等。类比较简单，读者可自行实现。

14 游戏开发实例

在本章，我们将实现一个飞行射击游戏。在飞行射击游戏中，很容易就能确立要建立的类：飞机类、子弹类、枪炮(子弹发射器)类。而无论是子弹还是飞机，都是飞行单位，有很多共同的属性，所以需要为其建立基类。

14.1 飞行对象基类

飞行对象基类需要处理飞行单位的移动，以及对边界做监测，飞行对象一旦移出屏幕则对游戏失去意义，将被设为死亡状态。而一旦处于死亡状态，则不再执行其更新方法。游戏中的飞行对象均处于友方和敌方两个阵营之中。

```
public class FlyObject {

    public enum GroupType
    {
        FRIEND,
        ENEMY,
    }

    private float border = 0;

    private Sprite sprite;
    private Vector2f pos;
    private Vector2f v;
    private Vector2f destPos;

    private boolean bDestPos;
    private boolean bLive = true;

    private float speed;

    public FlyObject(Vector2f pos, Vector2f velocity, Sprite sprite,float
speed) {
```

```
        this.pos = pos.clone();
        this.destPos = pos.clone();
        this.v = velocity.clone();
        this.sprite = sprite;
        this.sprite.setPos(new Size(this.pos.x,this.pos.y));
        bDestPos = false;
        this.speed = speed;
        border = Config.getInstance().screenSize.x/4;
    }

    public void setDestPos(Vector2f dest)
    {
        this.destPos = dest.clone();
        bDestPos = true;
    }

    public Vector2f getDestPos()
    {
        return destPos;
    }

    public void setSpeed(float speed)
    {
        this.speed = speed;
    }
    public void setPos(Vector2f pos) {
        this.pos = pos.clone();
    }

    public Vector2f getPos()
    {
        return pos;
    }

    public void setVelocity(Vector2f velocity) {
        this.v = velocity.clone();
    }

    public Vector2f getVelocity()
    {
        return v;
    }

    public void dead() {
        bLive = false;
        sprite.setVisible(false);
    }

    public void setSprite(Sprite sprite) {
```

```
        this.sprite = sprite;
    }

    public void reborn()
    {
        bLive = true;
        sprite.setVisible(true);
    }

    public boolean isLive()
    {
        return bLive;
    }

    public void update(long interval)
    {
        if(!bLive)
            return;
        if(bDestPos)
        {
            v = destPos.sub(pos).normalize().mul(speed);
            bDestPos = false;
        }

        interval% =20;

        pos.add(new Vector2f(v.x * interval, v.y * interval));

        sprite.setPos(new Size(pos.x,pos.y));

        if(pos.x < (0 - border) || pos.x > Config.getInstance().screenSize.x +
border || pos.y < (0 -border) || pos.y >Config.getInstance().screenSize.y +border)
        {
            dead();
        }
    }

    public Sprite getSprite()
    {
        return sprite;
    }
}
```

该类构造函数需要传入一个 sprite 对象，从而实现飞行对象与 sprite 精灵的绑定。每次更新会根据自身坐标来设置精灵的坐标。所有的飞行对象均可以设置目标位置，每次设置目标位置，会对自身的速度进行修正。当变成死亡状态时，绑定的精灵则变为不可见状态，减少渲染压力。

14.2 子弹和飞机的实现

14.2.1 子弹类

为了游戏更加丰富，子弹可以有多种类别，所以子弹类中添加一个枚举类型定义子弹种类，相比其飞行对象父类，子弹类增加了子弹类型和阵营两个成员变量，另外，在重生方法中根据阵营设置了子弹的不同颜色。

```
public class Bullet extends FlyObject {
    public enum BulletType
    {
        POINT,
        ELLIPSE,
    }
    private BulletType bType;
    private GroupType gType;
    public Bullet (Vector2f pos, Vector2f velocity, Sprite sprite,
BulletType bType,GroupType gType) {
        super(pos, velocity, sprite,0.1f);
        this.gType = gType;
        this.bType = bType;
        reborn();
    }
    public GroupType getGroupType()
    {
        return gType;
    }
    @ Override
    public void reborn()
    {
        super.reborn();
        if(gType == FlyObject.GroupType.ENEMY)
        {
            this.getSprite().setColor(1.0f,0.0f,0.0f,1.0f);
```

```
        }else
        {
            this.getSprite().setColor(1.0f,1.0f,1.0f,1.0f);
        }
    }

    public void setGroupType(GroupType gType)
    {
        this.gType = gType;
    }

    public BulletType getBulletType()
    {
        return bType;
    }
}
```

14.2.2 飞机及枪炮类

飞机需要有发射子弹的功能，但是和子弹不同，子弹产生时就在屏幕可见范围内，直接就处于有效的状态，而所有的飞机都得从屏幕外飞入屏幕。对于玩家而言，需要一个认知的过程，所以为初始的飞机设定一个屏幕内的位置，只有到达指定位置，才能进行发射子弹以及其余功能的启动。对于发射子弹而言，由于考虑到飞机发射子弹可能不止一处，例如升级后，左右均有炮管，以及子弹发射的数量可能上升，因此将发射子弹的实现剥离出来，定义枪炮类来实现，而飞机根据需要附加若干枪炮对象即可。枪炮类Gun代码如下：

```
public class AirGun {
    private int shootInterval;
    private Bullet.BulletType bulletType;
    private int bulletLevel;
    private long noShootTime = 0 ;
    private Plane owner;
    private BulletManager bulletManager;
    public AirGun(Plane owner)
    {
        shootInterval = 1000;
        bulletLevel = 1 + owner.getGroupType().ordinal()* 10;
        bulletType = Bullet.BulletType.POINT;
        noShootTime = shootInterval;
        this.owner = owner;
    }
```

```
    public AirGun (Plane owner, Bullet.BulletType bType, int level, int
shootInterval)
    {
        this.shootInterval = shootInterval;
        this.bulletLevel = level;
        bulletType = bType;
        noShootTime = 0;
        this.owner = owner;
    }

    public void shoot(long interval,BulletManager bulletManager)
    {
        noShootTime += interval;
        if(noShootTime < shootInterval)
        {
            return;
        }
        noShootTime = 0;

        switch (bulletLevel)
        {
            case 1:
                bulletManager.getBullet(owner.getPos().clone(), new
Vector2f(0,0.1f),bulletType,owner.getGroupType());
                break;
            case 2:
                bulletManager.getBullet(new Vector2f(owner.getPos().x+5,
owner.getPos().y),new Vector2f(0,10),bulletType,owner.getGroupType());
                bulletManager.getBullet(new Vector2f(owner.getPos().x-5,
owner.getPos().y),new Vector2f(0,10),bulletType,owner.getGroupType());
                break;
            case 3:
                bulletManager.getBullet(owner.getPos().clone(), new
Vector2f(0,10),bulletType,owner.getGroupType());
                bulletManager.getBullet(owner.getPos().clone(), new
Vector2f(-3,10),bulletType,owner.getGroupType());
                bulletManager.getBullet(owner.getPos().clone(), new
Vector2f(3,10),bulletType,owner.getGroupType());
                break;
            case 4:
```

```
                    bulletManager.getBullet(new Vector2f(owner.getPos().x+5,
owner.getPos().y),new Vector2f(0,10),bulletType,owner.getGroupType());
                    bulletManager.getBullet(new Vector2f(owner.getPos().x-5,
owner.getPos().y),new Vector2f(0,10),bulletType,owner.getGroupType());
                    bulletManager.getBullet (owner.getPos ().clone (), new
Vector2f(-3,10),bulletType,owner.getGroupType());
                    bulletManager.getBullet (owner.getPos ().clone (), new
Vector2f(3,10),bulletType,owner.getGroupType());
                    break;
                case 11:
                    bulletManager.getBullet (owner.getPos ().clone (), new
Vector2f(0, -0.15f),bulletType,owner.getGroupType());
                    break;
            }
        }
}
```

该类负责发射子弹，由于子弹有阵营属性，因此枪炮发射子弹时根据自身所属飞机阵营来决定子弹阵营。发射时，根据子弹级别实现不同的发射方式。

而飞机类则在创建时安装枪炮对象，由一个枪炮数组成员管理所安装枪炮，发射子弹时调用所有枪炮成员发射。

```
public class Plane extends FlyObject {

    public enum PLANE_COLOR {
        RED(1.0f, 0.0f, 0.0f, 1.0f),
        ORANGE(1.0f, 0.5f, 0.1f, 1.0f),
        YELLOW(1.0f, 1.0f, 0.0f, 1.0f),
        GREEN(0.0f, 1.0f, 0.0f, 1.0f),
        QING(0.0f, 1.0f, 1.0f, 1.0f),
        ZI(1.0f, 0.0f, 1.0f, 1.0f),
        GOLD(0.8f, 0.8f, 0.1f, 1.0f);

        private PLANE_COLOR(float r,float g,float b,float a)
        {
            this.r = r;
            this.g = g;
            this.b = b;
            this.a = a;
```

```
        }

        public float r(){return r;}
        public float g(){return g;}
        public float b(){return b;}
        public float a(){return a;}
         private float r,g,b,a;
    }
    protected boolean bShoot = false;
    protected boolean bTakePlace = false;

    public enum PlaneType
    {
        HERO,
        EGG,
        SYRING,
    }

    private PlaneType pType;
    private GroupType type;

    private List<AirGun> guns = new ArrayList<AirGun>();

    public Plane(Vector2f pos, Vector2f destPos, Sprite sprite, PlaneType
pType,GroupType type) {

        super(pos, new Vector2f(0,0), sprite,0.05f);
        this.type = type;
        this.pType = pType;
        setDestPos(destPos);
        reset();
    }

    public void reset() {
        bShoot = false;
        bTakePlace = false;
    }

    public PlaneType getPlaneType()
    {
        return pType;
```

```
    }
    public GroupType getGroupType()
    {
        return type;
    }

    public void addGun(AirGun gun)
    {
        guns.add(gun);
    }

    public void shootLicense(boolean license)
    {
        bShoot = license;
    }

    public void update(long interval,BulletManager bulletManager)
    {
        super.update(interval);

        if(!this.isLive())
            return;

        if(bShoot) {
            for (AirGun gun : guns) {
                gun.shoot(interval,bulletManager);
            }
        }

        if(!bTakePlace)
        {
            if(getPos().sub(getDestPos()).length()<10)
            {
                bTakePlace = true;
                bShoot = true;
            }
        }
    }
}
```

在飞机类中，定义了颜色的枚举类型，考虑到网络模式下有多个主机，为了区分，通过设置不同颜色来达到这个目的。

敌方飞机的功能在 plane 类中完全实现，但是，对于主机而言，还需要控制移动。主机总是一开始从屏幕下方升起，在到达指定位置才能开始控制，并发射子弹，所以添加控制标识，并在到达指定位置后将控制标识设为许可。

```
public class HeroPlane extends Plane {
    private boolean bControl = false;
    public HeroPlane(Sprite sprite) {
        super (new Vector2f (0, 0 ), new Vector2f (0, 0 ), sprite,
PlaneType.HERO, GroupType.FRIEND);
    }

    public void setColor(PLANE_COLOR color)
    {
        this.getSprite().setColor(color.r(), color.g(), color.b(),
color.a());
    }
    @Override
    public void reset() {
        super.reset();
        bControl =false;
        this.setPos(new Vector2f(Config.getInstance().screenSize.x / 2,
-this.getSprite().getSize().height));
        this.setDestPos(new Vector2f(Config.getInstance().screenSize.x
/ 2,Config.getInstance().screenSize.y / 4));

    }

    public boolean isControl() {
        return bControl;
    }

    @Override
    public void update(long interval,BulletManager bulletManager)
    {
        super.update(interval,bulletManager);

        if(bTakePlace && !bControl)
        {
            bControl = true;
            setVelocity(new Vector2f(0,0));
        }
        if(bControl) {
            if (getPos().x < this.getSprite().getSize().width / 2 || getPos().
x > Config.getInstance().screenSize.x - this.getSprite().getSize().width / 2
                || getPos().y < this.getSprite().getSize().height / 2 || getPos().
y > Config.getInstance().screenSize.y - this.getSprite().getSize().height / 2) {
                    setVelocity(new Vector2f(0, 0));
            }
        }
    }
}
```

14.2.3 子弹、飞机管理类

无论是子弹还是飞机，在游戏中产生和消失的频率都很高，当然不能不断地构建和析构对象。参照对象池的设计，分别为子弹和飞机定义两个管理类，其中各自维护一个列表存放子弹和飞机，在需要产生新子弹/飞机时，先查看列表中是否有死亡的、符合条件的子弹/飞机，有则激活/重生它，然后重新设定其位置、阵营等属性，再次使用；而如果列表中没有符合条件的对象，则创建新的对象，并加入到列表中。除此之外，管理类还负责遍历列表中所有对象，通知其更新。两个管理器代码类似，其中飞机管理类代码如下：

```
public class PlaneManager {
    private List<Plane> planes = new ArrayList<Plane>();

    private GameState state;
    private Frame[][] frames;

    private int heroCount = 0;

    public PlaneManager(GameState state,Frame[][] frames)
    {
        this.state = state;
        this.frames = frames;
    }

    public Plane createEnemyPlane (Vector2f pos, Vector2f destPos,
Plane.PlaneType pType)
    {
        if(Plane.PlaneType.HERO == pType)
            return null;
        //由于图片资源尺寸不统一
        Sprite sprite = state.createSprite (Config.getInstance ().
screenSize.x/10,Config.getInstance ().screenSize.y/10);
        sprite.add(frames[pType.ordinal()]);
        Plane plane = new Plane (pos, destPos, sprite, pType,
FlyObject.GroupType.ENEMY);

        switch (pType)
        {
            case EGG:
                AirGun gun=new AirGun(plane, Bullet.BulletType.
POINT,11,15000);
```

```
                    plane.addGun(gun);
                    break;
                case SYRING:
                    break;
            }
            planes.add(plane);
            return plane;
        }

    public HeroPlane createFriendPlane(Plane.PlaneType pType)
    {
        Sprite sprite = state.createSprite (Config. getInstance().
screenSize.x/10,Config.getInstance().screenSize.y/15);
        sprite.add(frames[pType.ordinal()]);
        HeroPlane plane = new HeroPlane(sprite);
        plane.setColor(Plane.PLANE_COLOR.values()[heroCount % Plane.PLANE
_COLOR.values().length]);

        AirGun gun = new AirGun(plane);
        plane.addGun(gun);

        heroCount++;

        planes.add(plane);
        return plane;

    }

    public Plane getEnemy(Vector2f pos, Vector2f destPos, Plane.PlaneType
pType)
    {
        for(Plane plane:planes)
        {
            if(!plane.isLive() && plane.getPlaneType()==pType)
            {
                plane.reborn();
                plane.setPos(pos);
                plane.setDestPos(destPos);

                return plane;
            }
        }

        //no find idle bullet
        Plane plane = createEnemyPlane(pos, destPos, pType);
        planes.add(plane);
        return plane;
```

```
        }
        public Plane getPlane(int index)
        {
            if(index >= planes.size())
                return null;
            return planes.get(index);
        }

        public int getPlaneCount()
        {
            return planes.size();
        }

        public void update(long interval,BulletManager bulletManager)
        {
            for(Plane plane:planes)
            {
                plane.update(interval,bulletManager);
            }
        }
    }
```

在该类中提供了创建敌机和主机的方法，而 getEnemy 是为游戏逻辑提供新的敌机，主机不会频繁地创建，所以不需要。敌机创建根据类型不同为其配备不同的枪炮。

14.3 碰撞检测类

碰撞检测在不同游戏中有不同的精确度要求，如果要求较低，如本项目，可以采用矩形碰撞检测的方法。事实上，在 Sprite 类中已经提供该方法。

```
public boolean isCollision(Sprite destSprite)
{
    return spriteRect.intersect(destSprite.getRect());
}

public boolean isCollision(Point point)
{
    return spriteRect.contains(point.x,point.y);
}
```

两个重载方法，前者判断两个精灵是否重叠，后者判断某点是否在精灵矩形范围内。如果使用精灵的矩形范围碰撞，其实是判断其纹理矩形是否相交，所以纹理图有效图像占图片尺寸的比例越大就越精确。碰撞检测针对飞机和子弹之间，数据直接来源于飞机和子弹的两个管理器，所以该碰撞检测类只提供一个静态方法来完成碰撞检测。

```
static public int collision (BulletManager bulletManager, PlaneManager
planeManager)
{
    int killEnemyCount = 0;
    for(int i =0;i <bulletManager.getBulletCount();i ++)
    {
        Bullet bullet = bulletManager.getBullet(i);
        if(bullet ==null || !bullet.isLive())
        {
            continue;
        }

        for(int j =0;j <planeManager.getPlaneCount();j ++)
        {
            Plane plane = planeManager.getPlane(j);
            if(plane ==null || !plane.isLive())
            {
                continue;
            }

            if (bullet.getGroupType().ordinal() + plane.getGroupType()
.ordinal() == 1) {
                if (bullet.getSprite().isCollision(plane.getSprite())) {
                    bullet.dead();
                    if(plane.getPlaneType()! =Plane.PlaneType.HERO){
                        plane.dead();
                        killEnemyCount ++;
                    }else if(((HeroPlane)plane).isControl()){
                        plane.reborn();
                        plane.reset();
                    }
                }
            }
        }
    }
    return killEnemyCount;
}
```

14.4 游戏中具体状态

每个游戏根据设计都有自身的诸多状态，在本节中，一共阐述 4 个状态，加载、菜单、主场景、游戏结束。

14.4.1 加载状态

加载状态应该完成资源加载、网络连接等游戏的准备工作。为了降低玩家等待的焦虑感，也提供给它加载的信息，在该状态中实现一个进度条，与第 13 章的 SpriteButton 相似，进度条也是通过两个精灵实现的，一个实现进度条的背景，一个实现前景。背景精灵是不变化的，而前景精灵则随着加载进度不断改变尺寸。下面代码，只是模拟了加载的过程，进度条是根据时间来变化的。

```
public class LoadingState extends GameState{

    private Sprite backgroundSprite;
    private Sprite loadingBackSprite;
    private Sprite loadingFrontSprite;
    private Frame frame;
    private int tick = 0;
    private Rect loadingRect = new Rect();
    @ Override
    public void onEnter() {

        super.onEnter();
         tick = 0;
         frame = new Frame(ResManager.getInstance(),R.drawable.welcome);
         backgroundSprite = createSprite(screenSize.x/2,screenSize.y/2,
screenSize.x,screenSize.y);
         backgroundSprite.add(frame);
         loadingRect.set(screenSize.x/6,screenSize.y/5,screenSize.x* 5/6,
screenSize.y/6);
         loadingBackSprite = createSprite(
                 loadingRect.centerX(),
                 loadingRect.centerY(),
                 loadingRect.width(),
                 loadingRect.height());
         loadingBackSprite.setColor(0.3f, 0.3f, 0.3f, 1.0f);
         loadingFrontSprite = createSprite(0,0,0,0);
         loadingFrontSprite.setColor(1.0f, 1.0f, 0.3f, 0.8f);
```

```
            updateLoadingSprite(0);
        }

        private void updateLoadingSprite(int value)
        {
            loadingFrontSprite.setSize(new Size(loadingRect.width() * value /
100.0f, loadingRect.height()));
            loadingFrontSprite.setPos(new Size(loadingRect.left + loading-
FrontSprite.getSize().width/2,loadingRect.centerY()));
        }
        @Override
        public void onExit() {
            super.onExit();
        }
        @Override
        public void onBack() {
        }
        @Override
        public void onLeave() {
        }
        @Override
        public void update(long interval,GameStateManager gsm) {
            super.update(interval,gsm);
            tick += interval;
            if(tick > 3000)
            {
                gsm.setState(GameStateManager.STATE.MENU);
            }else
            {
                updateLoadingSprite(tick/30);
            }
        }

        @Override
        public void handleEvent(MotionEvent event) {

        }

        @Override
        public boolean isEnded() {
            return false;
        }
}
```

14.4.2 菜单状态

菜单状态中用到了 SpriteButton。该状态通常很简单，提供几个按钮，供玩家选择，根据其点击位置响应相应按钮代表的功能。

```
public class MenuState extends GameState{

    private final int SINGGAME_BTN =1;
    private final int JOINGAME_BTN = 2;
    private SpriteButton singleBtn;
    private SpriteButton joinBtn;

    private int enterGameFlag = 0;

    @ Override
    public void onEnter() {

        super.onEnter();
        Frame singleFrame = new Frame(ResManager.getInstance(),
R.drawable.singlebtn);
        Frame joinFrame = new Frame(ResManager.getInstance(),
R.drawable.joinbtn);

        Sprite  singleNormalSprite  =  createSprite  (screenSize.x/2,
screenSize.y/3,screenSize.x/4,screenSize.y/15);
        Sprite  singleTouchedSprite  =  createSprite  (screenSize.x/2,
screenSize.y/3,screenSize.x/4,screenSize.y/15);
        singleNormalSprite.add(singleFrame);
        singleTouchedSprite.add(singleFrame);
        singleNormalSprite.setColor(0.60f, 0.70f, 0.0f, 1.0f);
        singleTouchedSprite.setColor(1.0f,1.0f,0.0f,1.0f);

        Sprite  joinNormalSprite  =  createSprite  (screenSize.x/2,
screenSize.y* 2/3,screenSize.x/4,screenSize.y/15);
        Sprite  joinTouchedSprite  =  createSprite  (screenSize.x/2,
screenSize.y* 2/3,screenSize.x/4,screenSize.y/15);
        joinNormalSprite.add(joinFrame);
        joinTouchedSprite.add(joinFrame);
        joinNormalSprite.setColor(0.60f, 0.70f, 0.0f, 1.0f);
        joinTouchedSprite.setColor(1.0f,1.0f,0.0f,1.0f);

        singleBtn = new SpriteButton(this,SINGGAME_BTN,singleNormal
Sprite,singleTouchedSprite);
        joinBtn = new SpriteButton(this,JOINGAME_BTN,joinNormalSprite,
joinTouchedSprite);

    }
    @ Override
```

```
    public void onExit() {
        super.onExit();
    }
    @Override
    public void onBack() {
    }
    @Override
    public void onLeave() {
    }
    @Override
    public void update(long interval,GameStateManager gsm) {
        super.update(interval,gsm);
        switch (enterGameFlag )
        {
            case 1:
                Config.net = false;
                gsm.setState(GameStateManager.STATE.GAMEMAIN);
                break;
            case 2:
                Config.net = true;
                //start net thread,waiting for connect...
                //gsm.setState(GameStateManager.STATE.GAMEMAIN);
                break;
        }
    }

    @Override
    public void handleEvent(MotionEvent event) {
        joinBtn.handleEvent(event);
        singleBtn.handleEvent(event);
    }

    @Override
    public void handleBtnEvent(int btnIndex) {
        switch (btnIndex)
        {
            case SINGGAME_BTN:
                enterGameFlag = 1;
                break;
            case JOINGAME_BTN:
                enterGameFlag = 2;
                break;
        }
    }

    @Override
    public boolean isEnded() {
        return false;
    }
}
```

14.4.3 主场景状态

如果比较复杂的游戏，该状态通常处理比较多的代码，绝大部分的游戏逻辑都在该类中实现。对于飞行射击游戏而言，相对还算简单，要做的事情主要有四个：背景不断滚动，控制主机移动(子弹是自动发射)，检测并处理子弹飞机间的碰撞，创建新的敌机。

```
public class GameMainState extends GameState {
    private HeroPlane heroPlane;
    private GameBackGround gamebg;

    private Frame backFrame;
    private Frame[][] planeFrames = new Frame[3][];
    private Frame[][] bulletFrames = new Frame[2][];

    private BulletManager bulletManager;
    private PlaneManager planeManager;

    @ Override
    public void onEnter() {

        super. onEnter();

        backFrame =new Frame (ResManager. getInstance (),R. drawable. background);
        Sprite backSprite1 = createSprite (screenSize. x/2, screenSize. y/2,
screenSize. x,screenSize. y);
        backSprite1. add (backFrame);
        Sprite backSprite2 =createSprite (screenSize. x/2,screenSize. y* 3/2,
screenSize. x,screenSize. y);
        backSprite2. add (backFrame);
        gamebg = new GameBackGround (backSprite1,backSprite2);

        planeFrames[0] = new Frame[2];
        //友机的帧信息
        planeFrames [ 0 ] [ 0 ]  =  new  Frame  (ResManager. getInstance ( ),
R. drawable. hero1);
        planeFrames [ 0 ] [ 1 ]  =  new  Frame  (ResManager. getInstance ( ),
R. drawable. hero2);

        planeFrames[1] = new Frame[1];
        planeFrames [ 1 ] [ 0 ]  =  new  Frame  (ResManager. getInstance ( ),
R. drawable. enemy1);
```

```
        planeFrames[2] = new Frame[1];
        planeFrames[2][0] = new Frame(ResManager.getInstance(),
R.drawable.enemy2);

        bulletFrames[0] = new Frame[1];
        bulletFrames[0][0] = new Frame(ResManager.getInstance(),
R.drawable.bullet1);
        bulletFrames[1] = new Frame[1];
        bulletFrames[1][0] = new Frame(ResManager.getInstance(),
R.drawable.bullet2);

        bulletManager = new BulletManager(this,bulletFrames);

        planeManager = new PlaneManager(this,planeFrames);

        heroPlane = planeManager.createFriendPlane(Plane.PlaneType.
HERO);
    }

    public Plane createPlane(Vector2f pos, Vector2f velocity,
FlyObject.GroupType gtype)
    {
        return null;
    }
    @Override
    public void onExit() {
        super.onExit();
    }
    @Override
    public void onBack() {

    }

    @Override
    public void onLeave() {
    }
    @Override
    public void update(long interval,GameStateManager gsm) {
        super.update(interval, gsm);
        gamebg.Move(interval);
        bulletManager.update(interval);
        planeManager.update(interval,bulletManager);
```

```
        CheckCollision.collision(bulletManager, planeManager);

        if(heroPlane.isControl())
        {
            if(Math.round(Math.random()* 50)% 50 == 1)
            {
                float posx = (float) (screenSize.x* Math.random());
                float destx = (float) (posx + Math.random()* 4 -2);
                planeManager.getEnemy(new Vector2f(posx, screenSize.y
+10), new Vector2f(destx, screenSize.y), Plane.PlaneType.values()[((int)
(Math.random() * 2 % 2 + 1))]);
            }
        }
    }

    @ Override
    public void handleEvent(MotionEvent event) {

        if(event.getAction() == MotionEvent.ACTION_DOWN)
        {
            int x = (int)event.getX();
            int y =Config.getInstance().screenSize.y - (int)event.getY();

            if(heroPlane.isControl())
            {
                heroPlane.setDestPos(new Vector2f(x,y));
            }
        }

    }

    @ Override
    public boolean isEnded() {
        return false;
    }
}
```

对于背景而言，通过同一个纹理创建两个精灵，然后循环拉动，创造无穷无尽的画面效果。

```
public class GameBackGround {
     private Sprite firstBgSprite;
     private Sprite secondBgSprite;
     private float speed;
     public GameBackGround(Sprite sp01,Sprite sp02)
     {
          firstBgSprite = sp01;
          secondBgSprite = sp02;
          speed = 0.1f;
     }

     public void Move(long interval)
     {
          firstBgSprite.setPos(new Size(firstBgSprite.getPos().width,
firstBgSprite.getPos().height - speed * interval));

          secondBgSprite.setPos(new Size(secondBgSprite.getPos().width,
secondBgSprite.getPos().height - speed * interval));

          if(firstBgSprite.getPos().height < - Config.getInstance().
screenSize.y/2)
{
          firstBgSprite.setPos(new Size(firstBgSprite.getPos().width,
secondBgSprite.getPos().height+Config.getInstance().screenSize.y));
          }
          if(secondBgSprite.getPos().height < - Config.getInstance().
screenSize.y/2)
          {
                 secondBgSprite.setPos(new Size(secondBgSprite.getPos().
width,firstBgSprite.getPos().height+Config.getInstance().screenSize.y));
          }
     }

     public void SetSpeed(float s)
     {
          speed=s;
     }
}
```

14.4.4 游戏结束状态

游戏结束状态，应该显示分数，甚至排行榜，然后提供一个按钮返回到主菜单或重新开始游戏。此处实现与游戏主菜单类似。

14.5 游戏的数据

游戏中子弹和飞机的信息记录各占一个表，而玩家的信息也需要一个表，所以在SQLite 中共创建三个表，子弹信息如表 14 - 1 所示，飞机信息表如 14 - 2 所示，玩家信息如表 14 - 3 所示。

表 14 - 1 子弹信息

_ID	name	level
1	Bullet1	1
2	Bullet2	1

表 14 - 2 飞机信息

_ID	name	gType	bullet
1	enemy1	1	1
2	enemy2	1	1
3	Hero1	0	2

表 14 - 3 玩家信息

_ID	name	level	planeId	planeColor
1	jeremy	1	3	0

相应的三个合同类的定义如下：

```
public class BulletContract {
    public static final String TABLE_NAME = "bullets";

    public static final String SQL_CREATE_TABLE = "CREATE TABLE "
            + TABLE_NAME + "("
            + BulletEntry._ID + " INTEGER PRIMARY KEY AUTOINCREMENT,"
            + BulletEntry.COL_NAME + " TEXT,"
            + BulletEntry.COL_LEVEL + " INTEGER);";
```

```
    public static abstract class BulletEntry implements BaseColumns {
    public static final String COL_NAME = "name";
    public static final String COL_LEVEL = "level";
    }
}
public class PlaneContract {
    public static final String TABLE_NAME = "planes";

    public static final String SQL_CREATE_TABLE = "CREATE TABLE "
            + TABLE_NAME + "("
            + PlaneEntry._ID + " INTEGER PRIMARY KEY AUTOINCREMENT,"
            + PlaneEntry.COL_NAME + " TEXT,"
            + PlaneEntry.COL_GTYPE + " INTEGER,"
            + PlaneEntry.COL_BULLET_ID + " INTEGER,"
            + " FOREIGN KEY (" + PlaneEntry.COL_BULLET_ID + ")
REFERENCES planes(" + BulletContract.BulletEntry._ID + "));";

    public static abstract class PlaneEntry implements BaseColumns {
        public static final String COL_NAME = "name";
        public static final String COL_GTYPE = "gtype";
        public static final String COL_BULLET_ID = "bullet";
    }
}
public class PlayerContract {
    public static final String TABLE_NAME = "players";

    public static final String SQL_CREATE_TABLE = "CREATE TABLE "
            + TABLE_NAME + "("
            + PlayerEntry._ID + " INTEGER PRIMARY KEY AUTOINCREMENT,"
            + PlayerEntry.COL_NAME + " TEXT,"
            + PlayerEntry.COL_LEVEL + " INTEGER,"
            + PlayerEntry.COL_PLANE_ID + " INTEGER,"
            + PlayerEntry.COL_PLANE_COLOR + " INTEGER,"
            + " FOREIGN KEY (" + PlayerEntry.COL_PLANE_ID + ")
REFERENCES planes(" + PlaneContract.PlaneEntry._ID + "));";


    public static abstract class PlayerEntry implements BaseColumns{
        public static final String COL_NAME = "name";
        public static final String COL_LEVEL = "level";
        public static final String COL_PLANE_ID = "planeId";
        public static final String COL_PLANE_COLOR = "planeColor";

    }
}
```

而对应数据库中的数据，定义相应的数据信息类来记录这些数据，包括 BulletInfo、PlaneInfo 以及 PlayerInfo。这些类比较简单，读者可自行实现。有了合同类和相应的数据信息类，可以在 SQLiteOpenHelper 的子类中添加对各种数据的读写方法：

```
public class DBHelper extends SQLiteOpenHelper{

    public static final int DB_VERSION = 1;

    public DBHelper(Context context) {
        super(context, GameConst.DataStorage.DB_NAME, null, DB_VERSION);
    }
    @ Override
    public void onCreate(SQLiteDatabase db) {
        try{
            db.execSQL(BulletContract.SQL_CREATE_TABLE);
            db.execSQL(PlaneContract.SQL_CREATE_TABLE);
            db.execSQL(PlayerContract.SQL_CREATE_TABLE);

            initDBData(db);
        }catch(Exception e){
            e.printStackTrace();
        }
    }

    @ Override
    public void onUpgrade (SQLiteDatabase db, int oldVersion, int
newVersion) {
        try{
            db.execSQL("DROP TABLE IF EXISTS" + PlayerContract.TABLE_NAME);
            db.execSQL(" DROP TABLE IF EXISTS" + PlaneContract.TABLE_NAME);
            db.execSQL("DROP TABLE IF EXISTS" + BulletContract.TABLE_NAME);
            onCreate(db);
        }catch(Exception e){
            e.printStackTrace();
        }
    }

    public void initDBData(SQLiteDatabase db)
    {
        ContentValues bulletValue = new ContentValues();
        bulletValue.put(BulletContract.BulletEntry.COL_NAME, "bullet1");
        bulletValue.put(BulletContract.BulletEntry.COL_LEVEL, 1);
        db.insert(BulletContract.TABLE_NAME, null, bulletValue);
```

```
        Cursor cursor = db.query(BulletContract.TABLE_NAME, null,
                BulletContract.BulletEntry.COL_NAME + "=?",
                new String[]{"bullet1"},
                null, null, null);
        cursor.moveToNext();

        int bulletID1 = cursor.getInt(cursor.getColumnIndex
(BulletContract.BulletEntry._ID));
        cursor.close();

        bulletValue.put(BulletContract.BulletEntry.COL_NAME, "
bullet2");
        bulletValue.put(BulletContract.BulletEntry.COL_LEVEL, 1);
        db.insert(BulletContract.TABLE_NAME, null, bulletValue);

        cursor = db.query(BulletContract.TABLE_NAME, null,
                BulletContract.BulletEntry.COL_NAME + "=?",
                new String[]{"bullet2"},
                null, null, null);
        cursor.moveToNext();

        int bulletID2 = cursor.getInt(cursor.getColumnIndex
(BulletContract.BulletEntry._ID));
        cursor.close();

        ContentValues planeValue = new ContentValues();
        planeValue.put(PlaneContract.PlaneEntry.COL_NAME,"enemy1");
        planeValue.put(PlaneContract.PlaneEntry.COL_GTYPE, 1);
        planeValue.put(PlaneContract.PlaneEntry.COL_BULLET_ID,
bulletID1);
        db.insert(PlaneContract.TABLE_NAME, null, planeValue);

        cursor = db.query(PlaneContract.TABLE_NAME, null,
                PlaneContract.PlaneEntry.COL_NAME + "=?",
                new String[]{"enemy1"},
                null, null, null);
        cursor.moveToNext();
        int planeEnemyID1 = cursor.getInt(cursor.getColumnIndex
(PlaneContract.PlaneEntry._ID));
        cursor.close();

        planeValue.put(PlaneContract.PlaneEntry.COL_NAME, "enemy2");
```

```
        planeValue.put(PlaneContract.PlaneEntry.COL_GTYPE, 1);
        planeValue.put(PlaneContract.PlaneEntry.COL_BULLET_ID,
bulletID1);
        db.insert(PlaneContract.TABLE_NAME, null, planeValue);

        cursor = db.query(PlaneContract.TABLE_NAME, null,
                PlaneContract.PlaneEntry.COL_NAME + "=?",
                new String[]{"enemy2"},
                null, null, null);
        cursor.moveToNext();
        int planeEnemyID2 = cursor.getInt(cursor.getColumnIndex
(PlaneContract.PlaneEntry._ID));
        cursor.close();

        planeValue.put(PlaneContract.PlaneEntry.COL_NAME, "hero1");
        planeValue.put(PlaneContract.PlaneEntry.COL_GTYPE, 0);
        planeValue.put (PlaneContract.PlaneEntry.COL _ BULLET _ ID,
bulletID2);
        db.insert(PlaneContract.TABLE_NAME, null, planeValue);

        cursor = db.query(PlaneContract.TABLE_NAME, null,
                PlaneContract.PlaneEntry.COL_NAME + "=?",
                new String[]{"hero1"},
                null, null, null);
        cursor.moveToNext();
        int planeHeroID = cursor.getInt(cursor.getColumnIndex
(PlaneContract.PlaneEntry._ID));
        cursor.close();
        ContentValues playerValue = new ContentValues();
        playerValue.put(PlayerContract.PlayerEntry.COL_NAME,"jeremy");
        playerValue.put(PlayerContract.PlayerEntry.COL_LEVEL,1);
        playerValue.put(PlayerContract.PlayerEntry.COL_PLANE_ID,
planeHeroID);
        playerValue.put(PlayerContract.PlayerEntry.COL_PLANE_COLOR, 0);
        db.insert(PlayerContract.TABLE_NAME,null,playerValue);
    }

    public PlayerInfo getPlayerInfo(long id)
    {
        SQLiteDatabase db = getReadableDatabase();
```

```
        Cursor cursor = db. query(PlayerContract. TABLE_NAME,null,
                    PlayerContract. PlayerEntry. _ID + "=?",
                    new String[] {String. valueOf(id)},
                    null,null,null);
        cursor. moveToNext();
        PlayerInfo player = new PlayerInfo();
        player. id = cursor. getLong(cursor. getColumnIndex
(PlayerContract. PlayerEntry. _ID));
        player. level = cursor. getInt(cursor. getColumnIndex
(PlayerContract. PlayerEntry. COL_LEVEL));
        player. name = cursor. getString(cursor. getColumnIndex
(PlayerContract. PlayerEntry. COL_NAME));
        player. planeId = cursor. getInt(cursor. getColumnIndex
(PlayerContract. PlayerEntry. COL_PLANE_ID));
        player. planeColor = cursor. getInt(cursor. getColumnIndex
(PlayerContract. PlayerEntry. COL_PLANE_COLOR));

        cursor. close();
        return player;
    }
    public void delPlayer(int id)
    {
        SQLiteDatabase db = getWritableDatabase();
        db. delete(PlayerContract. TABLE_NAME, PlayerContract. PlayerEntry. _ID +
"=?",
                    new String[]{String. valueOf(id)});
    }
    public void updatePlayer(PlayerInfo player)
    {
        SQLiteDatabase db = getWritableDatabase();
        ContentValues playerValue = new ContentValues();
        playerValue. put(PlayerContract. PlayerEntry. COL_NAME,
player. name);
        playerValue. put(PlayerContract. PlayerEntry. COL_LEVEL,
player. level);
        playerValue. put(PlayerContract. PlayerEntry. COL_PLANE_ID,
player. planeId);

        playerValue. put(PlayerContract. PlayerEntry. COL_PLANE_COLOR,
player. planeColor);

        db. update(PlayerContract. TABLE_NAME, playerValue,
                    PlayerContract. PlayerEntry. _ID + "=?",
                    new String[]{String. valueOf(player. id)});
```

```
    }

    public PlaneInfo getPlaneInfo(long id)
    {
        SQLiteDatabase db = getReadableDatabase();
        Cursor cursor = db.query(PlaneContract.TABLE_NAME,null,
                PlaneContract.PlaneEntry._ID + "=?",
                new String[] {String.valueOf(id)},
                null,null,null);
        cursor.moveToNext();
        PlaneInfo plane = new PlaneInfo();
        plane.id =
cursor.getLong(cursor.getColumnIndex(PlaneContract.PlaneEntry._ID));
        plane.groupType =
cursor.getInt(cursor.getColumnIndex(PlaneContract.PlaneEntry.COL_GTYPE));
        plane.name =
cursor.getString(cursor.getColumnIndex(PlaneContract.PlaneEntry.COL_NAME));
        plane.bulletId =
cursor.getInt(cursor.getColumnIndex(PlaneContract.PlaneEntry.COL_BULLET_ID));

        cursor.close();
        return plane;
    }

    public BulletInfo getBulletInfo(long id)
    {
        SQLiteDatabase db = getReadableDatabase();
        Cursor cursor = db.query(BulletContract.TABLE_NAME,null,
                BulletContract.BulletEntry._ID + "=?",
                new String[] {String.valueOf(id)},
                null,null,null);
        cursor.moveToNext();
        BulletInfo bullet = new BulletInfo();
        bullet.id =
cursor.getLong(cursor.getColumnIndex(BulletContract.BulletEntry._ID));
        bullet.level =
cursor.getInt(cursor.getColumnIndex(BulletContract.BulletEntry.COL_LEVEL));
        bullet.name =
cursor.getString(cursor.getColumnIndex(BulletContract.BulletEntry.COL_NAME));
        cursor.close();
        return bullet;
```

```
    }

    public void readAllInfo()
    {
        SQLiteDatabase db = getReadableDatabase();
        BulletInfo.bulletInfos.clear();
        Cursor cursor =
db.query(BulletContract.TABLE_NAME,null,null,null,null,null,null);

        while(cursor.moveToNext())
        {
            BulletInfo bulletInfo = new BulletInfo();
            bulletInfo.id =
cursor.getLong(cursor.getColumnIndex(BulletContract.BulletEntry._ID));
            bulletInfo.level =
cursor.getInt(cursor.getColumnIndex(BulletContract.BulletEntry.COL_LEVEL));
            bulletInfo.name =
cursor.getString(cursor.getColumnIndex(BulletContract.BulletEntry.COL_NAME));
            BulletInfo.bulletInfos.add(bulletInfo);
        }
        cursor.close();

        PlaneInfo.planeInfos.clear();
        cursor = db.query(PlaneContract.TABLE_NAME, null, null, null,
null,null,null);

        while(cursor.moveToNext())
        {
            PlaneInfo planeInfo = new PlaneInfo();
            planeInfo.id =
cursor.getLong(cursor.getColumnIndex(PlaneContract.PlaneEntry._ID));
            planeInfo.groupType =
cursor.getInt(cursor.getColumnIndex(PlaneContract.PlaneEntry.COL_GTYPE));
            planeInfo.name =
cursor.getString(cursor.getColumnIndex(PlaneContract.PlaneEntry.COL_NAME));
            planeInfo.bulletId =
cursor.getInt(cursor.getColumnIndex(PlaneContract.PlaneEntry.COL_BULLET_ID));
            PlaneInfo.planeInfos.add(planeInfo);
        }
        cursor.close();

        PlayerInfo.playerInfos.clear();
        cursor = db.query(PlayerContract.TABLE_NAME, null, null, null,
null,null,null);
```

```
            while(cursor.moveToNext())
            {
                PlayerInfo playerInfo = new PlayerInfo();
                playerInfo.id =
cursor.getLong(cursor.getColumnIndex(PlayerContract.PlayerEntry._ID));
                playerInfo.level =
cursor.getInt(cursor.getColumnIndex(PlayerContract.PlayerEntry.COL_LEVEL));
                playerInfo.name =
cursor.getString(cursor.getColumnIndex(PlayerContract.PlayerEntry.COL_NAME));
                playerInfo.planeColor =
cursor.getInt(cursor.getColumnIndex(PlayerContract.PlayerEntry.COL_PLANE_COLOR));
                playerInfo.planeId =
cursor.getInt(cursor.getColumnIndex(PlayerContract.PlayerEntry.COL_PLANE_ID));
                PlayerInfo.playerInfos.add(playerInfo);
            }
            cursor.close();
        }
}
```

附　录

外包是指企业将一些非核心、次要的或辅助性的功能或业务外包给企业外部的专业机构，利用他们的专长和优势来提高企业整体的效率。而服务外包则是指以 IT 作为交互基础的服务，服务的成果通过互联网交互，广泛应用于 IT 服务、人力资源管理、金融、会计、客户服务、研发、产品设计等众多领域。

服务外包根据其业务范围又可以分为信息技术外包服务(ITO)、技术性业务流程外包服务(BPO)、技术性知识流程外包服务(KPO)三类。2010 年财政部、国家税务总局、商务部、科技部、国家发展改革委员会联合发布的《关于技术先进型服务企业有关税收政策问题的通知》(财税〔2010〕65 号)中指出了这三种服务外包业务的适用范围，如附表 1 ~附表 4 所示。

1. 信息技术外包服务(ITO)

附表 1　软件研发及外包

类　别	适 用 范 围
软件研发及开发服务	用于金融、政府、教育、制造业、零售、服务、能源、物流和交通、媒体、电信、公共事业和医疗卫生等行业，为用户的运营/生产/供应链/客户关系/人力资源和财务管理、计算机辅助设计/工程等业务进行软件开发，定制软件开发，嵌入式软件、套装软件开发，系统软件、开发软件测试等
软件技术服务	软件咨询、维护、培训、测试等技术性服务

附表 2　信息技术研发服务外包

类　别	适 用 范 围
集成电路设计	集成电路产品设计以及相关技术支持服务等
提供电子商务平台	为电子贸易服务提供信息平台等
测试平台	为软件和集成电路的开发运用提供测试平台

附表3　信息系统运营维护外包

类　别	适用范围
信息系统运营和维护服务	客户内部信息系统集成、网络管理、桌面管理与维护服务；信息工程、地理信息系统、远程维护等信息系统应用服务
基础信息技术服务	基础信息技术管理平台整合等基础信息技术服务（IT 基础设施管理、数据中心、托管中心、安全服务、通信服务等）

2. 技术性业务流程外包服务（BPO）

附表4　技术性业务流程外包

类　别	适用范围
企业业务流程设计服务	为客户企业提供内部管理、业务运作等流程设计服务
企业内部管理数据库服务	为客户企业提供后台管理，人力资源管理，财务、审计与税务管理，金融支付服务，医疗数据及其他内部管理业务的数据分析、数据挖掘、数据管理、数据使用的服务；承接客户专业数据处理、分析和整合服务
企业运营数据库服务	为客户企业提供技术研发服务，为企业经营、销售、产品售后服务提供应用客户分析、数据库管理等服务。主要包括金融服务业务、政务与教育业务、制造业务和生命科学、零售和批发与运输业务、卫生保健业务、通信与公共事业业务、呼叫中心等
企业供应链管理数据库服务	为客户提供采购、物流的整体方案设计及数据库服务

3. 技术性知识流程外包服务（KPO）

适用范围：知识产权研究、医药和生物技术研发和测试、产品技术研发、工业设计、分析学和数据挖掘、动漫及网游设计研发、教育课件研发、工程设计等领域。